Rethinking Plastics in Product Design

Rethinking Plastics in Product Design

A Guide to Sustainable Transitions for the Environmental Emergency

Geoff Isaac
University of Technology Sydney, Australia

BLOOMSBURY VISUAL ARTS
LONDON • NEW YORK • OXFORD • NEW DELHI • SYDNEY

BLOOMSBURY VISUAL ARTS
Bloomsbury Publishing Plc
50 Bedford Square, London, WC1B 3DP, UK
1385 Broadway, New York, NY 10018, USA
29 Earlsfort Terrace, Dublin 2, Ireland

BLOOMSBURY, BLOOMSBURY VISUAL ARTS and the Diana logo
are trademarks of Bloomsbury Publishing Plc

First published in Great Britain 2025

Cover design: Louise Dugdale
Cover image: *Bell Chair*, Konstantin Grcic for Magis, 2020. Courtesy Magis.

A catalogue record for this book is available from the British Library.

A catalog record for this book is available from the Library of Congress

ISBN: HB: 978-1-3504-4913-8
         PB: 978-1-3504-4912-1
       ePDF: 978-1-3504-4915-2
      eBook: 978-1-3504-4914-5

Typeset by Integra Software Services Pvt. Ltd.
Printed and bound in India

To find out more about our authors and books visit www.bloomsbury.com
and sign up for our newsletters.

Table of Contents

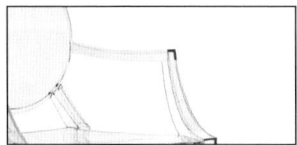

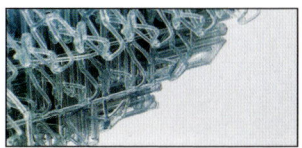

Conclusion 245

Figures

Introduction

Chapter 1

Chapter 2

Chapter 3

Chapter 4

Chapter 5

Chapter 6

Chapter 7

Chapter 8

Conclusion

Tables

Introduction

Chapter 4

Conclusion

Appendix

Acknowledgements

This book is based on research undertaken for my PhD dissertation. I would like to thank my supervisor, Distinguished Professor Peter McNeil, who encouraged me to enrol for the PhD programme at the University of Technology Sydney. I am extremely indebted to my co-supervisors, Dr Jesse Adams-Stein and Dr Stefan Lie. Jesse's timely and detailed feedback on my research and analysis has substantially contributed to this book. Stefan's experience as a practising industrial designer provided invaluable feedback on both the manuscript and the development of the ERPR tool.

Sylvia Katz was generous enough to read early drafts of Chapters 1 and 2 and provide feedback based on her expertise and experience from her distinguished career as a designer and author of several books on plastics. Nigel Howard, an expert life-cycle assessment (LCA) practitioner, reviewed the relevant section in Chapter 4 and provided valuable feedback for which I am grateful.

I would like to thank the following for their assistance and support: Jeffrey Meikle, Marcus Fairs (Dezeen), Angela Schad, Hagley Museum and Library (DuPont archive), Sanja and Steen Østergaard, Marjukka Törmi (Aarnio Design Ltd.), Virginia Marri (Archivo Gae Aultenti), Kathrin Hasskamp (Konstantin Grcic's office), Timothy Andreadis (Freemans), Emily Guthrie (Winterthur Museum), Trina Brown (National Museum of American History Library), Elizabeth Broman (Cooper Hewitt Library), Soufiane Bensabra, Virginia Wright, Louise Dennis (MoDIP), Kathy Hackett and Campbell Bickerstaff (MAAS), Lukas Holgersen (Thomas Pedersen's office), Mary Featherston; Alan Wells, Carolyn Clark and Susan Mossman (from the Plastics Historical Society); John Parrington, Karl Nethery and Troy Parker (from the Plastics Industry Manufacturers of Australia Inc; Maud Bury (Eugeni Quitllet's office), Jessica Pearson (Karim Rashid's office), Thales Melles (A Lot of Brasil), Steven Leclair (NRC Archives), Svein-Erik Hjerpbakk (Nordic Comfort Products AS), Therese Sanni and Leila Østerbø

(Snøhetta), Mathilde Wadoux (Starck Network), Phillipa Atkinson (Studio Dirk Vander Kooij) and Liliana Rodrigues (Marcel Wanders office).

I would also like to extend a special thanks to the designers and other industry participants who generously gave their time to participate in my research: Ron Arad, Dan Armstrong and Rene Linssen (Formswell), Edward Barber and Jay Osgerby, Gabriele Chiave (Marcel Wanders Office), Louis Durot, Tom Fereday, Manuel Garcia (Nagami), Sarah Gibson (DesgnByThem), Konstantin Grcic, Trent Jansen, Ander Lizaso (Iratzoki Lizaso), Ron Newman, Thomas Pedersen, Bertjan Pot, Tom Price, Eugeni Quitllet, Karim Rashid, Andrew Simpson (Vert Design), Philippe Starck, John Tree (Jasper Morrison), Gregg Buchbinder (CEO, Emeco), Ruben Hutschemaekers (then Head of Marketing and Communication, Magis), Kate Ringvall (then Sustainability Business Partner, IKEA), Mayda Diaz (Business Development & Technical Support, Bambacore), Victor Macadar (client of Bambacore) and Andreas Maegerlein (Group Leader Creation Center, BASF).

Finally, I would like to thank my partner, Robert Wellington, for his support, encouragement, and advice throughout my PhD studies and during the preparation of this book.

Abbreviations

ABS	Acrylonitrile Butadiene Styrene
AM	additive manufacturing
APCO	Australian Packaging Covenant Organisation
BPA	Bisphenol A
CAGR	compound annual growth rate
CCU	carbon capture and utilization
EPR	extended producer responsibility
ERPR	Environmentally Responsible Product Rating
ESG	environmental, social and governance
EU	European Union
Fibreglass	fibreglass reinforced plastic (usually epoxy or polyester)
GHG	greenhouse gas
GMOs	genetically modified organisms
HDPE	high density polyethylene
IPCC	Intergovernmental Panel on Climate Change
LCA	life-cycle assessment
MLP	multi-level perspective
MoMA	Museum of Modern Art, New York
NIMBY	not-in-my-backyard, resident action groups
OECD	Organisation for Economic Co-operation and Development
OPEC	Organization of the Petroleum Exporting Countries
PB	planetary boundaries
PC	polycarbonate
PE	polyethylene
PET	polyethylene terephthalate – a polyester thermoplastic
PFAS	per- and polyfluoroalkyl substances sometimes used as a grease and water repellent in food packaging
PHA	polyhydroxyalkanotes – polyesters produced in nature by processes such as bacterial fermentation of sugars

PHB	poly-β-hydroxybuturate (PHB), the first of the biodegradable plastics made through a fermentation process, often commonly referred to by the tradename Biopol
PLA	polylactic acid – a polyester thermoplastic derived from renewable resources
PMMA	poly(methyl methacrylate) acrylic glass, trademark names Plexiglas, Lucite or Perspex
PP	polypropylene
PS	polystyrene
PU	polyurethane
PVC	polyvinyl chloride
RIC	resin identification code
rPET	recycled PET
SUPs	single-use plastics
UN	United Nations
VCM	vinyl chloride monomer – an ingredient of PVC

Introduction

We knew it was coming and now here it is.[1]

When I began research for this book, in August 2021, wildfires burned across California, Canada, Greece, Italy, Turkey, Algeria, Lebanon, Cyprus, Russia, Bolivia, Brazil and Peru. Meanwhile, the United Kingdom, Germany, China, Japan, Turkey and Tennessee were grappling with the aftermath of devastating floods. The Intergovernmental Panel on Climate Change (IPCC) published a report detailing evidence to support its unequivocal claim that, 'human influence has warmed the atmosphere, ocean and land'. Concluding: 'many changes due to past and future greenhouse gas (GHG) emissions are irreversible for centuries to millennia, especially changes in the ocean, ice sheets and global sea level' (Masson-Delmotte et al., 2021).

Two years later, wildfires are so common that only the most devastating events make news headlines. Severe drought is impacting parts of North America, Europe, Asia and Africa while extreme rainfall has devastated other areas leaving thousands of people homeless in Australia, Japan, Puerto Rico and Libya while robbing tens of millions of their shelter and livelihoods in Pakistan, Nigeria and Chad. The negative impacts of climate change are now regularly being experienced in daily lives around the globe. 'We knew it [climate change] was coming and now here it is,' concluded one of the authors of the Intergovernmental Panel on Climate Change (IPCC) report. Most climate scientists agree, this is just the beginning, even worse and more frequent extreme weather events are on their way.

Meanwhile, the declaration of a climate emergency and pleas for net zero emissions are met with platitudes from a petrochemical industry committed to the continuous expansion of production to deliver 3–4 per cent more plastic every year through to 2050 or beyond (Nova Institute, 2020b, p. 18). At current growth rates, production of plastic will consume 15 per cent of the annual carbon budget by mid-century, making it impossible to reach global emission reduction targets (Ellen MacArthur Foundation, 2016, p. 17). Current industry growth projections are simply not compatible with a sustainable future.

Plastics are popular for their low cost, light weight, strength, durability, malleability, tactile warmth, impermeability, resilience to heat and chemicals, and the infinite colour range available. All of these attributes contribute to making plastics a favoured choice among product designers, material engineers, manufacturers and end-users alike. During the Covid-19 pandemic, plastics proved indispensable in personal protective equipment (PPE), the first line of defence against the virus. It is estimated we used 129 billion polypropylene (PP) disposable facemasks every month at the height of the pandemic, highlighting and exacerbating the ongoing waste problems caused by plastics (Xu & Ren, 2021).

As awareness of the environmental damage caused by both the creation and disposal of plastics increases, more urgent attention is being directed to finding alternative materials. But plastics are difficult, if not impossible, to replace for many applications, particularly in the health sector. More generally, substituting traditional materials like glass, steel or wood often fails to achieve superior environmental outcomes. Plastics, due to their prevalence, durability and cross-boundary effects present a 'wicked problem', lacking a clear objective optimal solution (Landon-Lane, 2018).

In response, there is growing interest in using recycled plastics or bioplastics (at least partially made from renewable biomass sources) for a growing range of applications. Many of these renewable carbon-based materials (renewable plastics) were originally conceived in efforts to curb the use of virgin fossil-based plastics (virgin plastics) in single-use packaging. Renewable plastics are now also being used to create consumer products, including chairs. I chose to focus on chairs as historically they have often been commissioned specifically to introduce or showcase new materials or technologies which have then been more widely adopted across many categories of industrial design. Many designers and manufacturers have already experimented with recycled plastics or bioplastics to make chairs. In total I examined sixty chairs made from renewable plastics and launched between 2007 and 2022 and these form the basis of the case studies referenced in this book. My analysis of these designs reveals how these new materials compare, in terms of their environmental impact, when used in real-life situations to make consumer products. More broadly, findings identify motivations and barriers for designers and manufacturers to experiment with renewable plastics.

The efforts of those involved with developing chairs using renewable plastics is commendable and deserves recognition. There is no intention to denigrate well-meaning work. However, without investigating and reporting the environmental impact of these innovations, future experimentation cannot benefit from, and build upon, the knowledge and experiences gained through these pioneering efforts. Designers and manufacturers committed to improving the environmental profile of their work need practical guidance in their daily decision-making. A clear vision and understanding of the necessary steps to achieve sustainability is vital for designers who want to be part of the transition to a more circular economy. Similarly, many end-users want to be confident their purchases are supporting products that deliver sound environmental outcomes and that they are not simply victims of 'greenwashing', designed to promote feelings of righteousness. This book addresses these everyday dilemmas, by sharing groundbreaking experiments with renewable plastics and quantitatively comparing their environmental impacts.

Between March 2019 and August 2021, I interviewed designers and manufacturers working with renewable plastics to gain insights into their motivations. Exerts from twenty-four of those interviews are used throughout this book (Table 0.1). I was fortunate to gain access to prominent industrial designers who have built international reputations for their work with market-leading manufacturers. These include Konstantin Grcic, Karim Rashid and Philippe Starck, together with representatives from manufacturers Magis and Emeco. Qualitative data obtained from those interviews supplemented the quantitative data discussed in Chapter 4. It should be noted that significant time has passed since these interviews were conducted and it is important to emphasize that opinions expressed at the time of interview might well have changed over time and not be an accurate reflection of the research participants' current beliefs.

Table 0.1 Interviews conducted with designers and manufacturers

Interviews completed	
NAME	**ORGANIZATION**
Designers (19):	
Ron Arad	
Dan Armstrong & Rene Linssen	Formswell
Edward Barber & Jay Osgerby	Barber Osgerby
Gabriele Chiave	Marcel Wanders
NAME	**ORGANIZATION**
Louis Durot	
Tom Fereday	
Manuel Garcia	Nagami
Sarah Gibson	DesignByThem
Konstantin Grcic	
Trent Jansen	
Ander Lizaso	Iratzoki Lizaso
Victor Macadar	Bambacore client
Thomas Pedersen	
Bertjan Pot	
Tom Price	
Eugeni Quitllet	
Karim Rashid	
Philippe Starck	
John Tree	Jasper Morrison
Furniture manufacturers (3):	
Gregg Buchbinder (CEO)	Emeco
Ruben Hutschemaekers Head of Marketing and Communication	Magis
Kate Ringvall Sustainability Business Partner	IKEA (Australia)
Material manufacturers (2):	
Mayda Diaz Business Development & Technical Support	Bambacore
Andreas Maegerlein Group Leader Creation Center	BASF

1. Why focus on chairs?

Chairs are central to our daily existence. Throughout our lives, as designer Galen Cranz observed, we spend most of our time seated: in different rooms in our home, in the office, on public transport or in cars, bars, cafes, restaurants or cinemas (Cranz, 1998, p. 15). While the basic design of chairs (four legs, a back and a seat) has remained largely unchanged over millenniums, they have been the subject of relentless attention from manufacturers and designers seeking to develop unique or distinctive versions. Manufacturers, keen to maximize profits, continually invest in new designs to service an increasing variety of niche markets or to appeal to the latest fashions and trends.

The brief for a chair is extremely challenging, requiring engineering and design skills. Chairs must be strong, appealing to look at, reasonably light, comfortable and fit for purpose. Chairs are often characterized as representing the quintessential design challenge. Anna Ferrieri, co-founder of Kartell and a designer herself, emphasized the engineering challenge; chairs, 'suffer the greatest structural strain a designer is likely to encounter' (Morello & Ferrieri, 1988, p. 138). Despite, or perhaps because of, these challenges, designers (with the support of their clients) continually seek to reinvent the chair, launching numerous new designs every year. Media scholar Ethan Zuckerman bluntly explained the allure of the challenge:

> Well, the chair has this really interesting place for the designers. Every designer wants to make a chair, but chairs are a fucking pain in the ass. They are really, really difficult. The back has to curve, it has to be slanted at a certain angle, and making it comfortable for people to sit on is a pretty serious challenge—that's why designers' chairs are signature artworks.
>
> (Suzdaltsev, 2015)

While Zuckerman's comments highlight the allure of the design challenge presented by a chair, German design duo Vogt + Weizenegger acknowledge the significant impact of technological innovations on the development of chairs:

> A chair design always illustrates the current status of society and its technological achievements (by combining) know-how with materials and aesthetic sentiments to form a seating sculpture linked to man's cultural history like no other item might.
>
> (Byars & d'Altoé, 2006, p. 156)

Vogt + Weizenegger were not the first to note the impact of technology on the development of chairs with George Nelson, who served as director of design for Herman Miller (from 1947 to 1972), made a similar observation in 1953: 'Every truly original idea – every innovation in design, every new application of materials, every technical innovation for furniture – seems to find its most important expression in a chair' (Nelson, 1994). As will be shown in Chapters 1 and 2, this observation applies to developments in the plastic chair market. Successful chair designs that withstand daily use and abuse act as envoys for the materials and manufacturing technologies used in their production.

Plastic chairs are most often associated with the cheap, mass-produced monoblocs (chairs manufactured in one process using a single material) which have flooded the market since the 1980s. Indeed, the market for plastic seating is projected to grow by 7 per cent (compound annual growth rate – CAGR) through to 2030, due to increasing demand for low-priced designer furniture (Grand View Research, 2022b). This growth rate is double the 3.5 per cent predicted for the entire plastics market, emphasizing the significance of this niche segment. But focusing on inexpensive monoblocs ignores a long history of high-quality, well-designed plastic chairs developed by some of the world's leading product designers. Designers have a long history of using plastics to design chairs for every conceivable market segment; from the bottom-end of the mass market to the upper echelons of high-end design collectors, where a luxurious, handcrafted chair made from fake (acrylic) glass and fake flowers (plastic roses) can pass between owners for more than a quarter of a million pounds (Figure 0.1).[2]

Plastics revolutionized the design and manufacturing process for chairs. With an estimated 60,000 plastics available (Huda & Bulpett, 2012, p. 338), designers can choose from a wide range of materials with different physical and aesthetic properties or can even specify the properties required and have one engineered to their specific needs. During the past eighty years, product designers have experimented with plastics in many forms, enabling them to introduce curvilinear shapes to seating solutions previously unobtainable using traditional manufacturing techniques and materials such as wood and metals. Indeed, chairs are often developed specifically to showcase a new material or technology, with practitioners keen to explore what they can bring to the ultimate design challenge, testing the boundaries of design and manufacturing capabilities.

The development of chairs specifically to showcase new materials or technologies highlights the importance of chairs as design objects and their role in shaping the public's perception of plastic as a material. In 2007, BASF held a competi-

Figure 0.1 *Miss Blanche* featuring plastic roses painstakingly entombed in acrylic resin. Designed by Shiro Kuramata (1934–1991), 1988 (edition of 56). Courtesy Sotheby's.

tion to design a chair to demonstrate the potential of using Ultradore High Speed plastic (polybutylene terephthalate – PBT) for applications beyond the automobile market, where the material was already well established. Konstantin Grcic (b.1965) won the competition and went on to produce the *Myto* cantilevered chair (Figure 0.2). UK-based industrial designers Barber & Osgerby were commissioned by Emeco to design the *On & On Stacking Chair* (2019) with a plastic comprising 70 per cent recycled polyethylene terephthalate (PET) derived from waste plastic bottles that would otherwise end up in landfill (Figure 5.24). A Lot of Brasil approached industrial designer Karim Rashid (b.1960) to develop the *Siamese Chair* using a plastic injection of the Amazonian fruit Acai and Ipe Roxo, probably the first bioplastic chair, launched in 2014 (Figure 0.3).

Chairs have been designed by people drawn from a variety of professions; interior designers, industrial designers and architects among them. In this book the generic term 'designer' has been used to refer to the person who designed a chair. Where specific reference is made to an individual designer, I have used their preferred professional title. The design of a chair is usually attributed to an

Figure 0.2 *Myto,* by Konstantin Grcic, Plank, 2008. BASF commissioned a cantilevered chair in an effort to develop a wider market for PBT, a material originally developed for the automobile market. Courtesy PLANK Srl.

individual 'heroic' designer (or design duo) responsible for conceptualizing the aesthetic and functional solution to the object. This practice masks the contribution made by specialist engineers, model makers and the entire team of professionals involved with the development of a chair. This 'fiction' of the 'designer genius at work in the studio' is becoming more difficult to sustain in today's era of distributed agency and expertise, as noted by anthropologist Arturo Escobar (Escobar, 2018, p. 85). Nevertheless, the practice of acknowledging the contribution of the designer over other participants continues.

Focusing on plastic chairs, and monoblocs in particular, affords the opportunity to examine separately the material and the artefact, and to interrogate that shifting relationship between them over time. We use hundreds, if not thousands, of plastic products every day, but many of those products include a mix of other materials, introducing complexity to the material-consumer relationship and limiting our analyses of material-object relationships. The study of plastic chairs offers a lens through which to understand the impact of technology and materials on design and the wider cultural and social implications of our relationship with plastic over time.

Perhaps most importantly, the capacity of a material to endure the mechanical stress and strains inherent in a chair demonstrates its viability for diverse

Figure 0.3 *Siamese Chair*, Karim Rashid, 2014. Courtesy Karim Rashid Inc for A Lot of Brasil.

applications within product design. Successful chair designs that incorporate new materials act as case studies showcasing the resilience of the materials used.

2. Sustainable transitions

While there is an increasing body of work focusing on design for transition, few studies have focused on how such change actually manifests. What does it really take to shift design and manufacturing practices, at scale, across complex supply chains? I use case studies of plastic chairs to examine the decision-making roles undertaken by in-house or independent designers and manufacturers when guiding everyday design decisions that impact environmental outcomes. As such I attempt to go 'beyond the canon', by considering the role of all actors representing technology, innovation, design and commerce as their roles entwine while developing new products. I identify strategies that progressive front-line actors can adopt toward more sustainable design solutions and the role they can play in transforming or reconfiguring the existing socio-technical system away from its dependence on fossil fuels.

We must all confront the fact that continually increasing the production of virgin fossil plastic is undesirable and unsustainable, even threatening our very existence. Designers and manufacturers have the agency and the ethical responsibility to accelerate the transition to more sustainable alternatives to virgin plastics by promoting the uptake of renewable plastics. In fact, I will demonstrate that those working with plastics to make consumer goods are in a powerful position to alter the course of the entire petrochemical industry. I aim to provide a blueprint to assist designers working with plastics to design a more sustainable future.

3. Structure of the book

The first section of this book (Chapters 1 and 2) provides a brief history of the development of the plastics market. Developments in the plastic chair market reflect are used to highlight innovations in the plastics industry and provide a context to map the shifting cultural attitudes toward the material. A key driver for developments in the plastic chair market during the twentieth century was dematerialization: reducing the amount of energy and/or material needed to manufacture a chair. Dematerialization was primarily motivated by industry's focus on maximizing profits but happily coincided with contemporary environmental imperatives to maximize resource efficiency.

Efforts to dematerialize chairs were initially rewarded with significant economic and environmental returns; however, those returns quickly diminished. At the same time, quality issues around early plastics began to emerge, coinciding with mounting concerns around the environmental and health impacts caused by the material. The initial allure of sleek, lustrous plastic surfaces, typical of the miracle materials emerging from science laboratories in the second half of the twentieth century, quickly dulled. These shifting cultural attitudes are mapped in Chapter 2.

In Chapter 3, the plastics industry is analysed with a particular focus on the growing environmental impacts implicit in current growth projections. A scenario or vision for a more suitable future for the industry is proposed. Chapter 4 explains the development of an eco-audit tool (ERPR) to enable a comparison of recent chairs designed using renewable plastics. Applying the tool enables the environmental profiles of the renewable plastic chairs to be examined. Chapter 5 identifies key lessons from this analysis that can provide guidance to all those seeking to reduce the environmental impacts when working with alternatives to virgin plastics.

Chapter 6 introduces a theoretical model (the muti-level perspective model – MLP) and applies it to the petrochemical industry with a specific focus on the use of plastics in consumer goods. This chapter examines the current operating environment and the impact of global events such as Covid-19 and the ongoing war in Europe on the industry, exploring how such disruptions to the status quo can create opportunities for less environmentally destructive technologies to gain acceptance.

Chapter 7 examines the barriers confronting both manufacturers and designers seeking to work with renewable plastics. The final chapter (8) identifies strategies and tactics that can be used by designers and manufacturers to accelerate the take up of renewable plastics. The conclusion itemizes the responsibilities for the those working with plastics with a checklist for those interested in developing an environmentally 'good' design with plastic.

1   Plastics, how we got in this mess

At the end of the nineteenth century elephants faced extinction. Nearly half a million kilograms of ivory was taken every year, mainly to supply the rapidly expanding market in Europe and the United States for billiards, the most sought-after entertainment pastime among society elites (Freinkel, 2011a, p. 15). Faced with the prospect of dwindling supplies, Phelan and Callender of New York, makers of ivory billiard balls, announced a competition to find an alternative to ivory (Smithsonian, 2012). The generous prize offered effectively kick-started research into synthetic plastics. Following six years of trial and error, John Hyatt produced a billiard ball made from celluloid (cellulose nitrate) in 1869. Unfortunately, this first semi-synthetic plastic proved unsatisfactory for the task; on impact the balls produced an explosive sound, like a gunshot (Freinkel, 2011a, p. 17; Friedel, 1983). Indeed, a Colorado saloonkeeper complained that guns were drawn in his bar 'every time the balls collided' (Freinkel, 2011a, p. 17).

Hyatt had created a substitute for a traditional material, but its first application proved unsuitable for use, helping to establish plastic's reputation as a cheap and inferior imitator. Quickly discarded, those billiard balls were soon joined by other early plastic items that had proved equally unsuitable for their intended purpose. Although often shoddily made, those rejected goods proved difficult to dispose of, exposing the most dangerous characteristic of fossil plastics, their endurance. Ever-growing quantities of discarded plastic products and packaging began to accumulate in landfills or litter our environment.

The introduction of plastics delivered an immediate reprieve for the elephants, but the true cost of their salvation remained uncalculated for over a century. Throughout the twentieth century, inexpensive, lightweight, fully synthetic plastics quickly became indispensable to daily life resulting in ever-increasing demand. It was to take nearly a century for the full environmental impacts caused by the creation and disposal of plastics to be exposed. Another fifty years later we are still struggling to find solutions to the problems created by plastics. But first let's take a step back and explore how and when we became addicted to plastics and examine the major technological developments that fed this addition and got us into the mess we find ourselves today.

1. Early plastics in design

Elephants were not the only species to benefit from the introduction of synthetic plastics. The Celluloid Manufacturing Company claimed, 'as petroleum came to the relief of the whale, [so] has celluloid given the elephant, the tortoise, and the coral insect a respite in their native haunts' (Friedel, 1983, p. 17). The chameleon-like ability of plastics to mimic other materials was exploited to make convincing, affordable versions of small objects traditionally made from ivory, bone, tortoise shell, hoof and horn among others. Expensive processes were invented so that plastics 'were introduced in the guise of substitutes, for although industry in the past has often feared and rejected originality it has welcomed substitutes', observed John Gloag in 1945 (Gloag, 1945, p. 17).

Historian Robert Friedel also claimed that the ability to convincingly imitate traditional materials was crucial to the acceptance of plastics by consumers (Friedel, 1983, p. 61). Disguised as the familiar, suspicions around the laboratory origins of the new materials were avoided. A material scientist suggested that 'by emphasizing the connection to ivory and tortoiseshell, manufacturers of celluloid were also communicating the material properties of the new, unfamiliar substance' (Madden, 2012, p. 8). Many products made from scarce natural resources like ivory or tortoiseshell were too expensive for the mass market. Convincing imitations of traditionally expensive luxury items were, for the first time, made available to people of moderate means.[1] British design academic Tom Fisher observed that the significant investment of time, skill and money needed to perfect the surfaces of moulds is reproduced perfectly on the surfaces of even the cheapest plastic objects, imbuing them with a sense of luxury (T. Fisher, 2015). However, imitating these natural materials to produce combs, buttons, knife handles, collar stays, dressing table and manicure sets also exposed the material to criticism; with many regarding plastic as the 'poor, ersatz cousin that was not quite as authentic or good as the original' (Madden, 2012, p. 8). French historian Bernadette Bensaude-Vincent argues the very fact that plastics could be used to imitate other materials meant the material was regarded with suspicion, its versatility and multipurpose nature causing unease (Bensaude-Vincent, 2013, p. 19). Bensaude-Vincent also highlights the concern that extremely light composites displayed the toughness, heat resistance and stiffness of steel – what could be more unnatural than that?

But plastics held an irresistible advantage; they could be moulded into every-day products of acceptable quality at a price point that appealed to a wider market, particularly relevant in helping to overcoming scarcities endured throughout the Second World War and igniting the consumer economy. The undeniable economic advantages of working with the material led to its rapid adoption by designers, manufacturers and consumers alike.[2] At the same time, these advantages were also regarded by others as proof that these new substances were no more than cheap substitutes, second-class or inferior materials, offering a merely 'plausible counterfeit of natural materials' (Boyd, 2011). They originated from alchemists' laboratories and not from nature; plastics were unnatural, lacking, 'social and physical authenticity' (T. Fisher, 2015, p. 126).

Design academic Ezio Manzini argues that for plastics to enter high culture they had to take on their own form and find their own image. Plastics became truly accepted once their economic value merged with cultural values. By showing their own qualities such as clarity or colourfulness, glossy surfaces, fluid forms and ability to be moulded in a single piece, plastics demonstrated that they were no longer imitating traditional materials. By not pretending to be anything other than what they are and by not denying their manufacturing processes, or apologizing for what they are, plastics were accepted, allowing them to 'take their place in the collective memory alongside existing materials' (Manzini, 2015, p. 54).

The plastics industry developed rapidly during the Second World War and in the decades following the end of the war the industry pivoted to feed the demands created from the emerging consumer society. Perhaps first to incorporate plastics in furniture was Charles Rennie Mackintosh who inlaid plastic panels into the surface of a smoker's cabinet in 1916 (Figure 1.1). In the late 1920s the Simmons Company in Racine, Wisconsin, produced five prototypes of an armchair made from Bakelite, but despite investing $50,000 to develop the eight moulds needed, the project was abandoned due to the chairs being excessively heavy (DuBois, 1972, pp. 180–181). Plastics historian Andrea DiNoto notes that by the 1930s designers began experimenting with plastics in furniture design using synthetic

Figure 1.1 Smoker's Cabinet, Charles Rennie Mackintosh, 1916. Bold yellow Erinoid triangles inlayed to contrast with the black painted woodwork. Courtesy © Victoria and Albert Museum, London.

(celluloid-based) lacquers, and laminates such as Formica for decoration (DiNoto & Arky, 1984, p. 193; Katz, 1978, pp. 12–13).

a. Acrylic glass

Rohm and Haas (now part of the Dow Chemical Company) developed poly (methyl methacrylate) (PMMA), commonly known as acrylic glass, in 1936, while researching polymer adhesives and branding their offering Plexiglas. Acrylic glass offers several significant benefits over silicate glass. It is less than half as dense, making products lighter and delivering savings during transportation through lower fuel consumption (Ashby & Johnson, 2014, p. 261). It also has good impact strength making acrylic glass ideal for experimentation across a wide range of consumer goods and as a structural material in furniture, such as chairs.

Grosfeld House, a New York retailer active from the 1930s to the 1950s, commissioned designers specializing in the Hollywood Regency style to create acrylic glass furniture which was showcased at the 1939 New York World Fair (Figure 1.2). Originating in the West Coast at the end of the 1920s, Hollywood Regency was

Figure 1.2 Acrylic glass chair, most likely by Lorin Jackson for Grosfeld House, 1939. Courtesy Josh Gaddy for Liz O'Brien.

based on neo-classical forms such as Adamesque and Biedermeier furniture and was quickly embraced by New Yorkers for its glitz and glamour. The translucent creations, commissioned by the company using acrylic glass, perfectly complemented the luxurious, silk, shearling and leopard print upholstery typical of the Hollywood Regency style. Plexiglas chairs quickly found support from the influential New York interior designer Elsie de Wolfe (1859–1950) who specified them for her clients. Meanwhile, in California, Swedlow Plastics, best known for manufacturing aircraft windshields, used Lucite (DuPont's brand of acrylic glass) to develop a range of transparent furniture including chairs from at least 1939, when they featured as props in a promotional film for DuPont (Figure 1.3).

Transparent acrylic glass can be used as a substitute for silicate glass. However, its strength, combined with its ability to be moulded, make acrylic glass suitable for a far wider range of applications, demonstrating the value of the synthetic over the natural material. DuPont capitalized on these strengths in an attempt to differentiate their acrylic glass (Lucite) from both silicate glass and Plexiglas and to develop post-war applications for the material. A 1940 sales brochure asserted that Lucite 'has been acclaimed, not as an imitation of something else, but as a material

Figure 1.3 Still from a 1939 promotional film for DuPont showcasing furniture made from Lucite, including the chair shown in the right-hand bottom corner. Courtesy Hagley Museum and Library, Wilmington, DE 19807.

capable of standing upon its own merits' ('Design for Modern Living', 1940, p. 7). The same issue of the company magazine featured a three-page article illustrating the wide applications to which Lucite had already been applied across all aspects of interior design, including furniture (Figure 1.4). Clearly it would not be practical to make most of these items with silicate glass. Acrylic glass opened new possibilities, quickly exploited by designers developing products, including chairs, which proudly celebrated plastic-as-plastic rather than an imitator of other materials.

New-York-based cosmetics mogul Helena Rubenstein (1872–1965) commissioned Hungarian artist Ladislas Medgyes (1892–1952) to design plastic chairs as part of a furniture suite for her city penthouse. These 1941 designs are often considered the first plastic chairs, even though both Grosfeld House and Swedlow Plastics had already produced designs from at least two years earlier. As the manufacturer of the Rubenstein chairs was the inventor of Plexiglas, it is almost certain these chairs are not made from Lucite, as is often reported.

All these early plastic chair designs were influenced by the Hollywood Regency style. Rejecting other styles popular at the time, such as Streamlining or Art Deco,

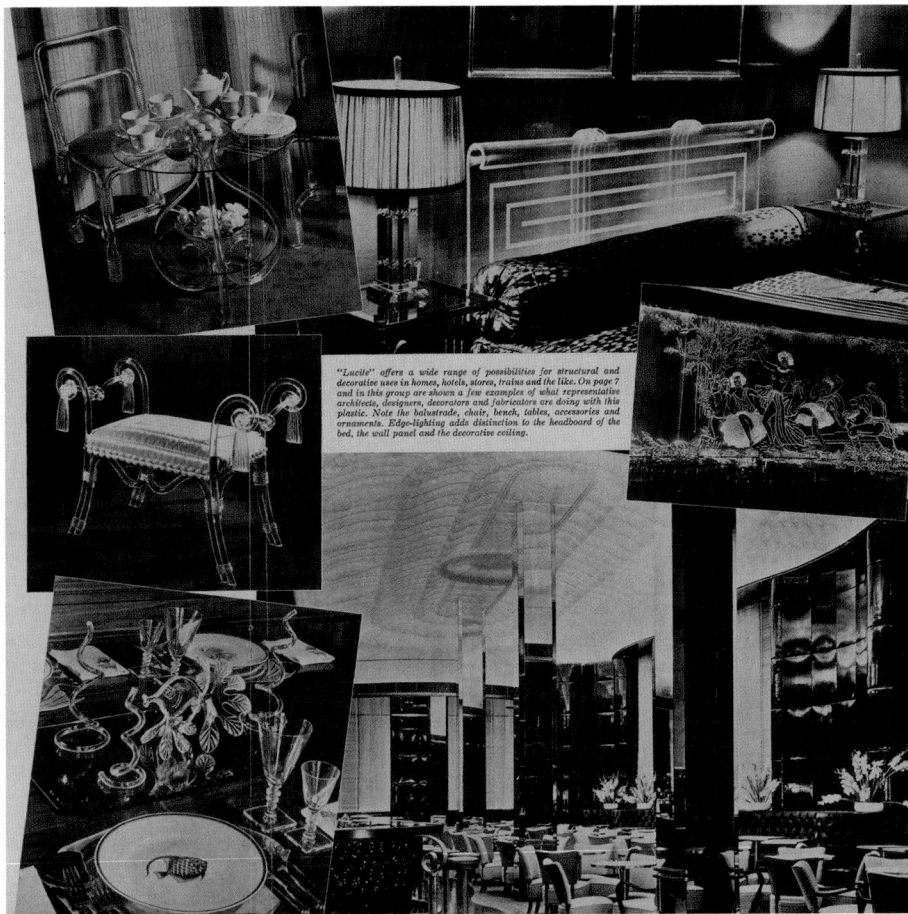

Figure 1.4 *DuPont Magazine*, April 1940 (pp. 8 and 9) showing various applications of Lucite across all aspects of interior design. Although not identified in the article, the chair, stool, lamp and side table can be attributed to LA-based Swedlow Plastics as featured in the company's 1941 brochure (Swedlow, 1941).[3] Courtesy Hagley Museum and Library, Wilmington, DE 19807.

these designs proudly feature ornate embellishments and decorative features, most clearly seen in the floral etchings on the chairs designed for Helena Ruben-stein (Figure 1.5) (DiNoto & Arky, 1984, p. 32). Replicating this style attracted criticism. British industrial design historian John Gloag claimed the designs 'were destitute of original inspiration' but conceding 'true they were unusual, but only in the way Cinderella's glass slipper was unusual' (Gloag, 1945). Gloag objected to the use of plastic as a direct substitute for traditional materials. Plastic had, by this time, earned a reputation for imitating more costly materials and these chairs reinforced this perception (DiNoto & Arky, 1984, p. 28). Gloag likely saw these early acrylic glass experiments as another example of plastic being used as a substitute, with the designs failing to exploit the true potential of the material, despite the claims made by DuPont.

Exploring the full potential of the new material was, however, limited by some practical constraints. The glass transition temperature of acrylic glass is above 100° Celsius, inhibiting hands-on experimentation.[4] In addition, the material was typically available only as precast sheets, rods or tubes; well suited to replicating the traditional components of chairs but unlikely to inspire the investigation of new forms. These factors guided designers to continue using traditional manufacturing techniques when working with acrylic glass. Perhaps, then, it is unsurprising that these early designs took the traditional approach to chair design and featured legs, seat and back as separate components joined together using long established manual techniques.

Significantly, these acrylic glass chairs were not aimed at the mass market commonly associated with most early plastic products but priced as to be obtainable only by the most affluent. News of these developments quickly reached Australia, with a January 1940 edition of the *Mercury* reporting, 'chairs and tables of a crystal-like plastic material, transparent yet strong, have appeared to astonish as well as please the home decorator looking for new ideas' ('Crystal-Like

Figure 1.5 *Helena Rubinstein Suite*, Ladislas Medgyes, 1941. Courtesy Wright (auctioneers).

Furniture', 1940). The article went on to concede that 'their cost as yet gives them the fashionable exclusiveness of very fine furnishings'. Plastic had begun to develop a dual reputation, cheap and poorly made products continued to flood the market but the material was increasingly being used in high-end products designed for the wealthy. An example of the phenomenon known as the plastic paradox.

Writing at this time, Victor Yarsley and Edward Couzens famously predicted that plastics held the promise of 'a new, brighter, cleaner and more beautiful world' (Yarsley & Couzens, 1941, p. 158). They saw a future where plastic would revolutionize everything from our homes and workplaces to the products we consume, and the clothing we wear. The book concludes with a utopian vision of the future predicting:

> This plastic man will come into a world of colour and bright shining surfaces where childish hands find nothing to break, no sharp edges, or corners to cut or graze, no crevices to harbour dirt or germs … The walls of his nursery … all his toys, his cot, the moulded light perambulator in which he takes the air
>
> (Yarsley & Couzens, 1941, p. 149)

The authors then describe in detail (over three pages) how plastics will be incorporated into every aspect of life until the very end:

> [In his old age the plastic man] wears a denture with silent plastic teeth and spectacles with plastic lens … until at last he sinks into his grave in a hygienically enclosed plastic coffin.
>
> (Yarsley & Couzens, 1941, p. 152)

Yarsley and Couzens' prediction is often quoted, but it is not unique, with many observers sharing similar visions. Five years earlier, even a regional newspaper in outback Australia published its own utopian prophecy – with an August 1935 edition claiming:

> Soon everybody will be sitting on plastic chairs, sleeping in plastic beds, eating food from plastic dishes set on plastic tables.
>
> ('All Plastic Home on Way', 1935)

By 1942, the local vision had been expanded with the *Woman's Mirror* claiming:

> You will soon rise from your plastic bed, bathe in your plastic bath, use your plastic brush and comb, and breakfast from a plastic table while

seated on a plastic chair. Later on you will pick up your plastic mounted handbag and hop into your plastic bodied car on your weekly shopping jaunt.

(Bacon, 1942)

These predictions were circulating throughout the Western world, but their utopian promises were interrupted by the outbreak of the Second World War, which was to have profound consequences for the development of the plastics industry. During the war, most industries were solely dedicated to the war effort, apparently curtailing Swedlow Plastics' attempts at diversification from aircraft components.[5] By the time the United States joined the war (1941), Japan had invaded many of its neighbours, preventing allied forces from accessing their traditional sources of rubber, tin, silk and other essential commodities.[6] The plastics industry quickly expanded production and developed new variants as substitutes for these materials to feed the war machine.[7] Chemists learnt to design at the molecular level, creating strong and tough polymers, specifically engineered to suit an ever broadening variety of uses (J. Meikle, 1995, p. 126). Importantly, armed service personnel relied on plastics to protect or even save their lives, earning a new respect for the material. Mortar fuses, parachutes, aircraft components, antenna housing, bazooka barrels, enclosures for gun turrets and helmet liners were among the countless products service personnel learnt to depend on ('Plastics in War', 1951).

The opportunity to gain an appreciation of the true potential of plastics extended beyond those serving on the front line. At the end of the Second World War, John Gloag predicted:

Women ... who have helped to build Spitfires and Hurricanes, are likely to welcome light and easily cleaned equipment in their houses, and especially in their kitchens; they are going to appreciate the smooth translucent and gaily coloured plastics that will be available for the making of kitchen equipment and furniture.

(Gloag, 1945, p. 41)

Gloag went on to claim that plastic will 'bring to the service of industry an array of new properties – new gifts of lightness, translucency, transparency, texture and colour' (Gloag, 1945, p. 17). He was right, demand soared. Following the end of the Second World War the industry quickly pivoted to satisfy unprecedented demand from a blossoming consumer society. In the United States alone, production of plastics reached 2.4 billion pounds in 1951, more than a tenfold increase from the 213 million pounds reported for the year the Second World War began (J. Meikle, 1995, p. 125).

Wartime innovations and applications demonstrated that plastics had potential well beyond their pre-war uses. Manufacturers realized that 'plastics have inherent advantage as structural materials in their own right, and that plastics often prove superior to other materials' ('Plastics in War', 1951, p. 9). In 1945, the inaugural issue of the local trade magazine, *Australian Plastics*, gave special attention to the material's use in the post-war home including plumbing, furniture and the vast scope of moulded plastics in the construction industry ('Australian Plastics', 1945). Together with government, industry was keen to reposition the material as war time uses were no longer in demand, resulting in excess supplies of plastic at a time when most traditional resources remained scarce. A year after the Second World War ended Sydney's *Daily Telegraph* bluntly announced:

> There's a lot of perspex [*sic*] about, because it was made for aeroplanes, which aren't in such big demand as they were a few years ago ... You've only got to look around the city shops, however, to see that it's going to be a plastic Christmas. Start in early with those guiding hints if you want to avoid receiving a transparent aeroplane part twisted into a toast rack, a coat-hanger, a set of swizzle sticks, or an occasional table.
>
> (Bartlett, 1946)

Faced with chronic shortages of most traditional materials, governments placed high expectations on the plastics industry to fulfil growing demands for housing and furniture from returning war veterans and pent-up demand for the many consumer goods unavailable during the war. Plastics displaced craft-based production, ushering in a new era of plenty that made an ever-increasing range of consumer goods accessible to households on even the most meagre budgets. The combination of modernist principles and the potential offered by plastics was even sufficiently powerful to transcend ideological barriers, with the newly formed Eastern Bloc embracing the industry to address the needs of its population following the end of the Second World War (Rubin, 2004).

Focusing attention on products rather than the material helped plastics to gain acceptance in many market segments, often by stealth. Plastics were rapidly integrated, almost unnoticed, into products designed to support every aspect of human endeavour. By the end of the 1940s, the new materials were fully entrenched in Western societies, including Australia, with an article in *Decoration and Glass* highlighting how plastics had already infiltrated every aspect of interior design and décor:

> 'What are plastics?' Your house and mine are full of these products; but because they are not all of one kind and frequently replace wood, bone or pottery, they often pass without recognition. We have come to accept them as part of everyday life without question ... [in] countless objects in daily use in every home.
>
> ('Modern Plastics', 1948)

This 1948 article illustrates that even during the Second World War, plastics had already made large inroads into every aspect of the daily life of Australians, displacing traditional materials in their wake. Plastics quietly slipped into the daily routine without significant advertising or editorial support from Australia's home-maker titles at that time (Figure 1.6). Among *Australian Home Beautiful*'s (*Home Beautiful*) editorial covering all aspects of home and garden wares, I found only one (short) mention of a plastic item, a nutcracker, in all the editions from 1945 to the end of that decade. There was no editorial mention of plastic products at all in *Australian House and Garden* (*House and Garden*). Plastic products, including chairs, were also notable for their absence from all images in both market-leading homemaker publications.

Despite the absence of advertising and editorial support, the ongoing short-ages of traditional materials continued to fuel demand for plastics. In 1950, the Melbourne newspaper *The Age* reported:

> The future of raw materials may well resolve itself into a major world problem. Only recently it was announced that the wool industry In the U.S.A. was rapidly falling off. Plastics may take the place of these threat-ened shortages. The world of tomorrow may be entirely dependent on plastic materials.
>
> ('The Age of Plastics', 1950)

At the same time it became increasingly apparent that many plastic products failed to live up to the utopian dream promised before the outbreak of the Second World War by Yarsley and Couzens and others.[8] The local trade magazine *Australian Plastics* reported on problems with a range of plastic products including 'riveted plastic baby harnesses which ripped apart, plastic patent handbags which lifted varnish from the counter, belts which stretched to ridiculous lengths, circular trays that warped and therefore wobbled'. The article went on to claim that people

Figure 1.6 During the 1940s the only plastic products being regularly advertised in Australia's homemaker magazine were: Rayon from Courtaulds (*Australian Home Beautiful*, November 1946), plastic light fittings from Duperite (October 1946), Nylex moulded products (December 1947) and Dulux (September 1949), who started promoting 'synthetic' paint toward the end of the decade. Author's collection.

had stopped buying plastic products to avoid further poor experiences ('Plastics Covered Chair', 1951, p. 9). These concerns were also expressed in the consumer press, with *House and Garden* reporting:

> You may have bought a mixing bowl because someone has told you that 'plastic is unbreakable'. Then you are justifiably annoyed when it breaks the first time it is dropped, or when the tumblers you bought for the children withstand breakage, but melt when placed near the heat of the kitchen stove.
>
> (Wright, 1951, p. 25)

An Australian Government report included similar findings and noted that 'the general reputation of the industry has suffered in the past through particular instances of shoddy production' ('The Australian Plastics Industry', 1950). Plastic became increasingly associated with 'tacky' in all senses of the word: cultural, structural and sensorial (T. H. Fisher, 2004, p. 24). In Australia the plastics industry was comprised of many small producers, with little regulation or standards in place. As a result, inferior goods remained commonplace and continued to tarnish the industry's reputation. Disappointing consumer object-interactions shaped their material-object relations, negatively impacting the reputation of plastic (Shove, 2007, p. 110).

Despite those quality issues, the plastic home invasion continued. A 1952 newspaper article listed 'new' (synthetic) plastic versions of many commonly found household objects, including tableware, ashtrays, clock cases, lighting equipment, brushes, radio cabinets and cosmetic jars. The journalist observed that plastics were making their way into the kitchen and bathroom, being used for everything from aprons and tablecloths to even the kitchen sink. They concluded: 'laminated plastic veneers have come into extensive use as a building material and for furniture' ('Plastics Invade the Home', 1952).

Plastics began to represent specific cultural values. Advertising increasingly appeared from the early 1950s, frequently emphasizing the durable, hygienic, easy-to-clean and heat-resistant plastic surfaces, available in an unprecedented variety of colours, as the main benefits of plastic for homemakers (Figure 1.7). This extended to furniture, with *House and Garden* reporting that hard-wearing laminated kitchen furniture 'will not stain and can be wiped clean with a damp rag' (Wright, 1951, p. 76).

While plastics had gained a foothold in most markets by stealth, they made a more dramatic intrusion into the furnishing industry. Designers who introduced the material to the industry chose to celebrate the material, unashamedly displaying

Figure 1.7 Aristoc brochure featuring laminated top table with bent chrome chairs upholstered in vinyl, *c.* 1950. Author's collection.

brightly coloured laminates, transparent acrylic glass and, in particular, glass-fibre reinforced plastic (fibreglass) which encouraged the exploration of exciting new forms.

b. Fibreglass

Unlike acrylic glass, fibreglass enabled designers to use their artisanal craft and sculpting skills to experiment using their own hands. In the simplest 'hand lay-up' process, glass fibres (glass blown into fine stands) are spread in a mould by hand and brushed over with epoxy or polyester resin at room temperature. Additional layers can be added to reinforce as required, giving a new level of control over the design's strength at any point while facilitating economical use of the material.[9] Once the resin cures, the solidified moulded item can be removed, trimmed and polished. The process facilitates the production of complex curvilinear shapes. When working with fibreglass 'the industrial designer was someone completely involved in the manufacturing process, from initial conception and drawing, handling of materials, to final production stages' (Collins, 1988, p. 28). Designers could materialize their ideas for plastic chairs all the way from the drawing board to the drawing room. Richard Schultz (1926–2021), an industrial designer working with Knoll from 1956 to 1972, reported that during his early years with the company:

> Knoll was a very simple company, no department of design and develop-
> ment, so you had to do everything yourself. You designed the furniture,
> you invented the way to make it, you designed the jigs and the tooling, and
> for a little while you were in charge of production.
>
> (Lutz et al., 2012, p. 117)

The design process at this time typically progressed from hand drawings to scale models and finally full-sized models, with each stage testing the artisan skills of the designer and their teams. While this process slowed progress it created the significant advantage of time to reflect. Designers could (literally and metaphorically) live with drawings, models and prototypes, using the time to re-examine, reconsider, refine and revise every aspect of the creation to perfect every line, join and curve.

Moulds could be made from inexpensive traditional materials like wood or plaster, helping to minimize the upfront financial investment. Design historian Penny Sparke argued that plastics like fibreglass 'discouraged a reductive approach towards design and encouraged, instead, an expansion of possibilities' (Sparke et al., 1993, p. 103). Designers could break free from the restriction imposed by traditional construction techniques (Katz, 1985, p. 14). Cultural historian Jeffrey Meikle claimed that the introduction of fibreglass to the furniture industry freed design from 'a "cubist" or "constructivist" project of assembling discrete parts, as in the early modern movement' (J. Meikle, 1995, p. 198). Designers could explore extreme organic, compound and curvilinear forms that had previously eluded them.

The sleek, futuristic appearance of fibreglass, with its polished, glossy, reflective, warm-to-the-touch surface, embodied the utopian ideals circulating during the Space Age (1957–1969). Writing on the history of the chair, Galen Cranz argued:

> Designers in the twentieth century, starting with the proponents of Art
> Nouveau, were interested in finding ways to express their own age, so
> they did not want to use historical styles.
>
> (Cranz, 1998, p. 80)

Finding 'ways to express their own age' was primarily facilitated by the availability of new materials and construction techniques and chief among these was glass reinforced polyester.

Chairs from this period differed from most traditional seating solutions as they featured rounded edges rather than sharp-edged corners. The main driver for their appearance was purely practical, as the material flows better in the mould when edges are curved. However, the production process delivered a seductive aesthetic as well (J. Meikle, 1995, pp. 115–116). The curves reflect light from at least one highlight when viewed from any angle, delivering a glint, or blink, to catch the eye. Curves are still preferred by consumers according to a study which found that rounded shape products are considered cosier, sexier, more elegant and less masculine when compared with the materials of sharp-edged products. Rounded plastic is perceived as more futuristic than its sharp-edged counterpart (Karana, 2010, p. 278).

All the advantages offered by the material made fibreglass ideal for independent designers and small businesses, resulting in its widespread adoption, quickly displacing acrylic glass. Fibreglass enabled Charles and Ray Eames (1907–1978 and 1912–1988) to finally develop the one-piece shell chair first envisaged by Charles and Eero Saarinen for their winning entry in the *Organic Design Competition* organized by MoMA in 1940 (Noyes, 1941, pp. 1, 12–17). Initial attempts to develop the design using plywood failed, and a decade passed before fibreglass provided the solution (Figure 1.8). Presented without upholstery, the armchair

Figure 1.8 *DAX chair* (Dining height Armchair X-Base), Charles & Ray Eames, Herman Miller Furniture Company, Zenith Plastics Co., 1950. Courtesy © Vitra Design Museum, photo Andreas Sütterlin.

proudly displayed the rough textured, semi-gloss finish of the fibreglass and was available in a limited colour range. With no springs or even padding, these chairs presented a real departure from the norm and were initially only accepted by the commercial market.

Eero Saarinen (1910–1961) disliked the raw finish featured on the Eames chairs and sought to find a technique to improve the surface for his *Tulip Chairs* (Figure 1.9) (Lutz et al., 2012, p. 170). Using veil matting with fine denier on the surface of the chair during compression moulding encouraged more resin absorption, creating a perfectly smooth skin, free from blemishes and sensual to touch. Compared to the Eames chairs, the finish achieved is smoother, exuding a promise of cleanliness and ease of maintenance. The additional investment needed to perfect this finish was immediately rewarded with commercial success in the residential market, something that had initially eluded the Eames design.[10]

The original plan to produce the *Tulip Chair* as a plastic monobloc was stymied when fibreglass proved unable to support the weight of a sitter. Saarinen solved this problem by casting the base in aluminium. Great attention was paid to

Figure 1.9 *Tulip Chair* (No. 151, left) and armchair (no. 150, right), Eero Saarinen for Knoll 1958. Courtesy © Vitra Design Museum, photo Andreas Sütterlin (left); Jürgen Hans (right).

matching the finish of the two surfaces and refining the curve and join between the shell and base, all to enhance the illusion of a monobloc. It was perhaps the first example of a product being disguised to look like plastic, reflecting growing consumer acceptance of synthetics.

Fibreglass, summarized Bensaude-Vincent, was originally developed for 'weight saving and cost reduction in transport and handing. However, [it] generated a deep change in design, and facilitated a new approach to materials research' (Bensaude-Vincent, 2013, p. 20). Designers could create both the material and the product in the same single process, with matter and form generated 'in one single gesture' (Bensaude-Vincent, 2013, p. 20). Ironically, it had taken the development of a totally synthetic material to enable the execution of the most extravagant, and elaborate, organic designs inspired by nature. The 1960s saw the introduction of numerous biomorphic chair designs featuring brightly coloured, highly polished, glossy surfaces. Fibreglass was plentiful and relatively inexpensive allowing designers to explore ideas without a need to conserve resources. Bensaude-Vincent observed that during the 1960s and into the 1970s, 'shining, fluorescent and flashy surfaces' (Bensaude-Vincent, 2013, p. 24) prevailed over the traditional preference for more natural-looking pastel colours across most product categories, and this applied to furniture (Figures 1.10–1.13) (Bensaude-Vincent, 2013, pp. 24–25).

Examples of fibreglass designs from the Space Age

Figure 1.10 *No. 412/Karuselli*, Yrjö Kukkapuro for Artek, 1965. Courtesy Artek.

Figure 1.11 *Elda*, Joe Colombo for Comfort, 1965. Courtesy © Vitra Design Museum.

Figure 1.12 *Pallo* or *Ball Chair*, Eero Aarnio, 1965. Courtesy Eero Aarnio's Archives.

Figure 1.13 *Culbuto Chair*, Marc Held for Knoll, 1972. Courtesy Knoll Archives.

Fibreglass demanded manual skills, creating the opportunity for a designer to be involved with all stages of production; however, this was also its main drawback. While models and moulds can be handcrafted for a relatively low cost, the manufacturing process is heavily dependent on manual work and hand finishing, making marginal costs high, ultimately restricting sales to an affluent niche market.[11]

In a publication on material selection product designers are encouraged to observe an important principle:

> People prefer products that are on the one hand maximally novel while being as familiar as possible … if you decide on a very novel shape, you may be well advised—from an aesthetic point of view—to stick to a familiar material for the product category.

> (Karana et al., 2013)

Many mid-century designers had the courage to ignore this advice, experimenting with both unfamiliar new materials and dramatic novel shapes simultaneously. By breaking the rules and pushing boundaries many of these designs achieved both critical and commercial success, with some designs destined to achieve iconic status and remain in production for decades. Many of the successful designs from that era display a chameleon-like ability to blend into any surroundings, looking equally at home in modern, post-modern, minimalist and period rooms.

Despite these long-term successes, the immediate impact of fibreglass furniture was limited, due to their expense. For instance, in 1971, Victor Papanek, a pioneering advocate of ecologically responsible design, claimed that due to their high price the real impact of the Eames chairs had been negligible (V. Papanek, 2019, p. 122). In Australia the furniture industry remained conservative, preferring to stick with designs with a proven track record (Featherston, 1971, p. 15). An executive from Hoechst Australia noted that 'even the very rich, generally approach their lifestyles with a caution of conservatism' (McDonald, 1973, p. 22). The executive went on to observe that plastic furniture was only 'purchased by a modish elite, who are drawn to the design dynamic', adding that avant-garde furniture would look 'ludicrously misplaced' in the average home.

While plastic furniture remained relatively scarce, this was not a symptom of any outright rejection of the material. By the time the first locally produced plastic chair appeared in 1956, consumption of plastic had reached 5 pounds per head of population per year in Australia and 'ranged from garden hoses, children's toys, tablecloths and curtains, glass reinforced plastic boats and car bodies' ('Plastics', 1956). Plastic components were also now commonplace in 'electrical switch gear, radio and TV circuits, telephone handsets, translucent corrugated sheeting, and self-lubricating bearings, and gears' ('Plastics', 1956).

By the end of the 1950s, any reservations regarding plastics were overshadowed by benefits as perceived by consumers in Australia, Europe and North America. Parents held sufficient faith in the material to allow their children to play with an increasing range of plastic toys. The Hula Hoop was introduced in 1957, followed by Lego in 1958 and Barbie in 1959. Even money went plastic, with American Express launching the first credit card in 1958. Plastics quickly 'went from being the stuff of imitation in the 1950s, to being valued in their own right as the materials of fashionable design' (T. Fisher, 2013, p. 287).

By the early 1960s, plastics had become common place across every aspect of Australian interior decoration. A local newspaper reported that the 'plastic invasion' had begun in the kitchen, spread to the bathroom and was now infiltrating the lounge, the bedroom and the remaining rooms of the house, even being used for outdoor furniture ('Plastics Give Luxury Living', 1962). Having exhausted original markets, the plastics industry searched for opportunities to diversify, adding more products to compete in more markets.

c. Injection moulding

Injection moulding pre-dates the development of synthetic plastics.[12] Plastic pellets are fed from a hopper, heated, then the molten plastic is forced under pressure into a mould. Once cooled, the moulded component is removed ready for use or requiring minimal finishing. When introduced to the industry, injection moulding radically altered the economics of producing furniture. In addition to the financial disruption, injection moulding distanced the relationship between the designer and the manufacturing process and shifted the end-users' relationship with their processions.

Injection moulding significantly altered the role of the designer. The skills of the designer, no matter how talented, were no longer sufficient to succeed alone or with a small team. Injection moulding demanded a network of skilled specialists and significant financial resources to develop moulds and secure access to the machinery needed for production. Designers could no longer rely on their own resources; they needed the support of a manufacturing partner with access to the necessary skills and infrastructure. Big business became attracted to the industry, keen to concentrate ownership of production facilities to maximize the economies of scale offered by the new technology.[13]

Plastic is the perfect material for moulding as it is homogeneous and free from defects, unlike natural materials like wood which is grained and knotted and varies in quality. Even metals, glass and ceramics can contain defects or variations in quality interrupting production. Polymers are isotropic, the consistency of physical properties ensuing production can continue uninterrupted (Morello & Ferrieri, 1988, p. 32). Acrylonitrile Butadiene Styrene (ABS), polypropylene (PP) and polyethylene (PE) enabled single-shot injection plastics to perform similarly to fibreglass, but at significantly reduced cost. Reinforced fillers, fibre mats, lacquers and laminates were no longer required, slashing production time (J. Meikle, 1995, p. 191).

Injection moulding is ideally suited to modernist principles, as it is much simpler and less expensive to create moulds devoid of ornamentation or decoration. As with compression moulding, curved edges are easier to produce than sharp-edged corners. Modernist principles, favouring sleek, clean lines and the elimination of decorative additions were also well aligned with industry's focus on maximizing profits. Economic drivers propelled injection moulding, ensuring rapid adoption by those involved with the production of consumer goods, including furniture.

Despite the many benefits offered by injection moulding, industrial designers were forced to adapt to the limitations of the technology. Moulds could no longer be made by hand using wood and plaster. Creating a mould in steel or aluminium demanded an increasingly specialized skill set which Gino Colombini described as sculpting in steel (Morello & Ferrieri, 1988, p. 45). To ensure success, designs for moulded components must be adapted to respond to feedback from experts on how the material will behave in the mould or where strengthening is necessary. To get into production a designer must hand over the project to a toolmaker. Once the mould is formed, the ability of the designer to make changes is limited, as any alterations are usually deemed prohibitively expensive.

Injection moulding made it impossible for designers to retain hands-on involvement with the entire process. French design theorist Raymond Guidot claimed that designers working in the early days of injection moulding often had the least opportunity for involvement with the production of their creations (Guidot, 2006, p. 278). Guidot's opinion is in line with Papanek's, who claimed that objects bearing the signs of their maker's hand create a 'spiritual link between the producer and the user', suggesting that, when using these production methods, the connection is lost (V. Papanek, 1985). The designer's artisan skills, so useful when working with fibreglass, became largely redundant once a component enters the pre-production stage. The designer can become alienated from the production process, their ability to continuously modify and improve a design, as was previously possible when working with fibreglass, is limited. In the digital age, designers benefit from greater control over their creations, keeping them involved for longer than when it was necessary to hand a design project over to the R&D or engineering departments (Guidot, 2006, p. 278). With both digital technology and fibreglass, designers can retain direct control over their creation from design to finished product, a level of involvement simply not possible with injection moulding.

Injection moulding dramatically transformed seating solutions, providing a new level of versatility and mobility. Lightweight, injection-moulded chairs were no longer confined to a fixed location. This contrasted with the then popular three-piece suite, heavy static objects often placed in a single location for months or years at a time. Seating arrangements became dynamic; chairs could be picked up and moved around at will – grouped together for quiet conversations or re-arranged in rows for wing.

d. Development of the monobloc

Injection moulding proved to be the ideal technology to deliver the modernist vision of 'machine furniture', minimizing manual labour inputs. The technology also proved essential to the successful production of chairs made using a single material using just one production process, known as monoblocs.

Monoblocs are attractive to designers and manufacturers for their potential to deliver saving through dematerialization; reducing the quantity of materials consumed, eliminating labour hungry assembly of components, reducing energy consumption during production and transportation (as less fuel is required to distribute lighter goods). Additionally, monoblocs can be designed to stack, further improving transport efficiencies. Dematerialization happily aligned industry's focus on maximizing profits with sustainability principles, as reduced demand for energy and resources at the individual product level promised to deliver lower environmental impacts.

Designers had experimented with wood and sheet metal to develop a chair from a single piece of material since at least the 1920s but to little avail.[14] A prototype for a monobloc plastic chair was created using compression moulding by Arthur Donahue (1917–1996) and Douglas Simpson (1916–1967) for the *Design in Industry* exhibition, held in Ottawa as early as 1946, but the prototype failed to progress to production (Figure 1.14) (Collins, 1988, p. 34). Indeed, another two

Figure 1.14 Design in Industry Exhibition – experimental chair produced by D. Simpson and A. J. Donahue of Winnipeg, 1946, of plastic made by the National Research Council. Courtesy Library and Archives Canada/National Film Board of Canada fonds/a160515.

decades would pass before a monobloc would be mass produced using injection moulding.

By definition, plastics can be moulded into almost any shape, making it no surprise that the prospect of creating a monobloc using plastic quickly became an obsession for many leading designers of the day. Design historian John Collins observed, 'this process [creating a monobloc] would be the aesthetic and technical goal of every designer during the post-war period' (Collins, 1988, p. 34). Grant Featherston (1922–1995) (an Australian mid-century designer active in furniture design from 1946 to 1975) held similar views and was inspired by the challenge to achieve structural unity:

> Producing the integral one-piece plastic chair stands at the pinnacle of the furniture designer's aspirations.
>
> ('Design Plays a Stellar Role in Exports', 1980, n.p.)

> The optimum design might be found in a product which comes complete from the mould, which for transport requires no wrapping, which also occupies a minimum of space – through its stackability [sic], foldability or volume reduction by compression.
>
> (Featherston, 1971, pp. 32–35)

Designers and manufacturers were also acutely aware that, in addition to the creative challenge, development of a monobloc held the prospect of significant financial rewards, due to lower production costs and the potential for mass-market sales. A durable design in plastic, suitable for both indoor and outdoor use, could appeal to multiple lucrative niches across both the commercial and residential markets. Additionally, the space savings that could be achieved by stackable monoblocs opened significant sales potential among venues with dynamic seating requirements.

Ultimate success in creating a monobloc required collaboration between a designer and a manufacturer. The designer (working with engineers) had to develop techniques to ensure the structural integrity of the design under load, while the manufacturer had to access or develop injection-moulding technology sufficiently advanced enough to produce objects as large as a chair. Easy solutions to either of these tasks were not to be found, and progress toward a monobloc was slow.

European designers were at the forefront of exploring the potential offered by injection moulding. Robin Day (1923–2000), Britain's most successful mid-century designer, achieved outstanding commercial success with *Polyside*, developed in collaboration with UK manufacturer Hille and launched in 1961 (Jackson, 2011, p. 122). Injection-moulded PP was used to form the shell, while steel was used to form a selection of bases for the chair (Figure 1.15). *Polyside* featured a textured surface to create a non-slip, matte surface in contrast to the glossy, shiny plastic-as-plastic finish typical of Space Age fibreglass chairs and the Italian designs of this period.

Figure 1.15 *Polyside*, Robin Day for Hille, 1963. Courtesy © Hille.

It was to be Italian designers and manufacturers, Kartell in particular, who then took the lead in advancing toward developing a monobloc. Kartell was founded in 1949, by the chemist Giulio Castelli (1920–2006) and his wife, architect Anna Castelli Ferrieri (1918–2006). This combination of technical expertise and design talent proved crucial for the company's success, with Castelli leading the polymer research while Ferrieri designing many commercially successful products for the company (Sparke et al., 1993, p. 84).

The *Empilable* was the first chair developed by Kartell using injection mould-ing (Figure 1.16). It took over four years and fifty prototypes to perfect the 1963 design and locate 'the point of coinciding – of the needs of the new technol-ogy to those of the design' (Morello & Ferrieri, 1988, p. 136). Richard Sapper (1932–2015) and Marco Zanuso's (1916–2001) child's chair was light, stackable

Figure 1.16 *Empilable 4999/5*, Richard Sapper and Marco Zanuso for Kartell, 1964. Courtesy Santi Caleca for Kartell Museum, Milan.

and available in a range of bright colours. Only a children's chair could be made due to limitations on the size of injection-moulding equipment available at the time (Máčel et al., 2008). Production of the *Empilable* still required the legs to be moulded separately and clipped individually into place below the seat.

Building upon the success of the *Empilable*, Joe Colombo (1930–1971) was challenged by Kartell to develop an adult version of the chair, launched in 1964 (Figure 1.17). The legs of the *Universale* were again moulded separately from the shell due to limitations of the size of machines.

Despite the commitment and investments made by Kartell, it was German designer Helmut Bätzner (1928–2010) in collaboration with manufacturer Bofinger who eventually, in 1967, succeeded in creating the first monobloc chair (Figure 1.18). Decades of research financed by German industry had led to significant advances in polymer adhesives, acrylic glass and injection-moulding machines. It is fitting, then, that despite the prolonged efforts of Italian designers, the first monobloc was developed in Germany. The *Bofinger BA 1171* (named in tribute to Rudolf Baresel-Bofinger, the owner of the manufacturer) became the prototype for the ubiquitous monobloc chair, of which countless millions have since been

Figure 1.17 *Universale* (No. 4867), Joe Colombo for Kartell, 1967. Courtesy Santi Caleca for Kartell Museum, Milan.

Figure 1.18 *Bofinger BA 1171*, Helmut Bätzner for Bofinger-Stuhl, 1966. Courtesy © Vitra Design Museum, photo Jürgen Hans.

made by numerous manufacturers all over the world. The original version took four minutes to create using the then new pre-impregnation compression moulding (prepreg) of fibreglass (Fiell & Fiell, 2009, p. 111). During two years in development, scrupulous attention led to the L-shaped, splayed legs with triangular feet, developed to maximize strength while minimizing the footprint. The efficiency of this design solution is best demonstrated by comparing the legs of the *Bofinger* with those of an inexpensive monobloc available from hardware stores today. The angle and shape have hardly changed in over fifty years, despite endless investment in computer technology (from CAD to AI), materials and manufacturing technologies.

The *Selene* (1969) was primarily inspired by German technology, not the efforts of Kartell, according to manufacturer *Artemide*'s founder, Ernesto Gismondi, who had seen Reglar (fibreglass reinforced polyester) being used to make photographers' developers trays during an overseas trip. These trays led Vico Magistretti (1920–2006) to design the *Selene*, a complex process that took nearly eight years to complete (Figure 1.19). Working with fibreglass sheets (later replaced with

Figure 1.19 *Selene*, Vico Magistretti for Artemide, 1969. Courtesy © Vitra Design Museum, photo Jürgen Hans.

ABS) just three mm thick to form the unique S-shaped legs allowed the design to retain an elegant simplicity while meeting the demands of regular use (*Vico Magistretti – Heller*, n.d.). Magistretti explained:

> I didn't want a chair to be composed of several different parts, I wanted a single unit; nor did I want to create a chair like Joe Colombo's [*Universale*] that, with its thick, heavy legs, looked like an elephant, although it was well designed and very imaginative.

(Pasca, 1991)

Development of a cantilevered chair had long intrigued designers and, in the 1960s two Danish designers raced to develop the first monobloc solution.[15] Steen Østergaard's (b. 1935) *290* (Figure 1.20) possibly pre-dates the *Panton Chair,* commonly considered the first cantilevered monobloc. The *290* features a sweeping curved base, giving more legroom while enhancing the gravity-defying illusion of the design. The chair may not have entered production until 1970, although, in emails to the author, the designer insists the *290* was available from 1966. Production of the *290* required the manufacturer, Cado, to invest in what they claimed to be the world's largest injection-moulding tool. Once in production a chair could be moulded in just 132 seconds. After being removed from the mould the chair had to be allowed to dry before the seams were sanded, varnished and packed for delivery. This design then still fell short of the ultimate goal, to design a chair that could be moulded in one-piece and emerge from the mould ready for use.

The S-shaped *Panton Chair* (Figure 1.21) is one of the best-known plastic chairs, nominated by Peter Fiell as 'one of the most important events in the entire history of furniture design' (Gosnell, 2004). Of all the Scandinavian designers Danish Vernon Panton's (1926–1998) work is the most synonymous with the Space Age movement. He used vibrant and exotic colours to create a futuristic world using the latest materials and manufacturing techniques for products and interiors that can be best described as funky or 'out there'.[16] Panton worked on the design for the S-shaped chair from 1956 and by the end of the decade had built a full-scale model of the stackable chair made from polystyrene (PS) foam. Panton did then create the first design for a one-piece moulded cantilevered chair but getting the design into production proved to be both problematical and time-consuming. Scandinavian manufacturers were unwilling to take on the challenge, but in 1962, Panton secured backing from Vitra. Five years and ten prototypes later, the *Panton Chair* launched in the Danish design journal *Mobilia*

Figure 1.20 *Cantilever 290*, Steen Østergaard for Cado1966/1970? For Cado, re-issued by Nielaus in 2013 (left). Courtesy Nielaus.

Figure 1.21 *Panton*, Verner Panton for Vitra, 1967 (right). Courtesy Vitra.

in August 1967, causing a sensation (despite the first series being restricted to just 150 copies).[17] In 1998, Vitra introduced PP to the production process, enabling a true monobloc (i.e. one process with no finishing required) version of the design to be finally produced (*Panton Chair – Verner Panton – Official*, 2018).

Engineer and entrepreneur Henry Massonnet (1922–2005) arguably perfected the monobloc with the *Fauteuil 300* in 1972. The chair was produced in a single injection-moulding process in less than two minutes (since reduced to one), with no hand finishing required (Figure 1.22). Massonnet's breakthrough design linked the front legs to the backrest using the armrests, minimizing the material required for support while enhancing stability. The chair flexes to distribute applied weight over all four legs, adjusting to support the sitter regardless of their position (Niermann, 2004). Using the armrests to enhance stability provided an engineering solution to the monobloc challenge yet to be surpassed. The monobloc also set a new benchmark in energy efficiency, with the reduced cycle time reflected in energy saving (Mianehrow & Abbasian, 2017).

During the 1980s, low-priced monoblocs based on this design started appearing globally but, in a tribute to the efficiency of Massonnet's solution, no visibly significant improvements to the basic design have been made. Virtually identical chairs are available from hardware stores or garden centres everywhere, with the design remaining almost unchanged over the past fifty years. Global availability

Figure 1.22 *Fauteuil 300*, Henry Massonnet, 1972. Courtesy © Vitra Design Museum, photo Jürgen Hans.

was aided by manufacturers in high-income countries replacing moulds when they begin to show signs of wear. Second-hand, slightly worn moulds are then sold for a fraction of their replacement cost to manufacturers in low- and middle-income earning countries, enabling even more cost-effective production and lower retail prices.[18] Unfortunately, quality can be compromised with the relentless pursuit of a lower price resulting in compromises on the quantity and quality of raw materials.

The entire process to injection mould a monobloc, measured in seconds, is far quicker than compression moulding and produces consistent results of a higher quality, thanks to the greater level of control offered by the process. Manual labour during production can be virtually eliminated, as only minimal (if any) hand finishing and packing are required, delivering significant economies of scale for furniture manufacturers.[19] After incurring the (very considerable) upfront expense of the moulds, the pure marginal cost of producing a component using this technology is relatively insignificant, little more than the cost of raw materials. The process is therefore ideally suited to mass production – the bigger the production run, the smaller the average cost per unit becomes, facilitating retail prices that are accessible to a broad audience.

The monobloc is divisive among designers and consumers alike. This can be explained by the dichotomy between its indisputable engineering success and its design credentials, often viewed as dubious. Critics argue the white monobloc has become ubiquitous and is seen as the ultimate symbol of global mass consumption homogenizing the world, leading to cross-cultural convergence, a definitive example of globalizing one-worldism (Fry & Nocek, 2021, p. 39). An image of a monobloc offers no clue about its location in time or space. Ethan Zuckerman, then at MIT, stated, 'I have a hard time thinking of other objects that are equally independent of context' (Zuckerman, 2011). However, Zuckerman also acknowledges the monobloc as a 'victory of high modernist design', suggesting the designer's goal is to achieve universal acceptance for their objects:

> They never want them to be only culturally specific. They want to transcend, so it can be used by all people. So maybe this is the high modernist design culture just on a cultural level where everybody can afford it.
>
> (Suzdaltsev, 2015)

While Zuckerman applauds the ubiquitous nature of the monobloc, regional variations do exist. While the white monobloc dominates high-income markets, bright primary colours are often favoured across Asia (Friedrichs & Eickhoff, 2010, p. 12). 'Red is a favourite amongst the Chinese, green is well-liked by the Malays and blue is preferred by Indians' (Zhuang, 2018). In Sri Lanka, monoblocs come in many colours and with ornate floral and flame-like designs. Malaysian manufacturer Mah Sing claims the market for their white chairs is limited to funeral parlours (Zhuang, 2018).

Smithsonian Magazine claimed the monobloc is 'in the worst possible taste, so cheap, ugly and everywhere'. The monobloc is accused of turning outdoor eating spaces in America 'into tawdry, second-rate imitations' of European style cafes (Gosnell, 2004). Indeed, some European cities (including Bern, Basel, Zurich, Bratislava and Barcelona) have gone as far as banning monoblocs from public spaces (Friedrichs & Eickhoff, 2010, pp. 14, 19). Hank Stuever, in the *Washington Post*, accused the stacking chair of being 'the Tupperware container of a lard-rumped universe' (Gosnell, 2004). Ralph Caplan (a design consultant) says the chair is a sad reflection of consumer society, satisfying the lowest common denominator with little regard for quality (Rybczynski, 2017). The monobloc is the ultimate incarnation of the plastic paradox, an engineering triumph that has made relatively comfortable seating available to billions of people across the planet for the first time. Dematerialization has delivered the modernist ideal of democratizing design, but that success has resulted in billions of plastic chairs flooding the market. The monobloc has homogenized the world, perhaps more so than any other single design object, but it is also responsible for a significant contribution to the environmental disaster caused by plastic waste. Monoblocs symbolize the complexity of material culture and they are the definitive symbol of globalized consumer modernity, while destined to become the most common future relic of the fossil plastic age.

2. Summary

The abundant availability of plastics, combined with the scarcity of traditional materials following the end of the Second World War, led to the rapid adoption of an ever-increasing variety of consumer goods made from plastic, many of which even arrived packaged in plastic. Plastics, and their increasingly mechanized production methods, fuelled the booming post-war consumer society.

Plastic technologies advanced rapidly in the second half of the twentieth century. Early plastics were suitable for processing by individuals or small companies using low-cost tools and machinery, but these were quickly superseded by materials best suited to mass production techniques requiring significant capital investment and attracting the interest of big business with a focus on profits. Arguably the most significant change in plastic production during this period was the introduction of injection moulding, the economics of which centralized production while making plastic products (including chairs) accessible to a much wider audience. Injection-moulding technology played a pivotal role in realizing the modernist vision of a 'machine furniture'. While many new materials and technologies were trialled by industry, the pursuit of profit ensured that those which successfully dematerialized a product by delivering saving in energy or resource use were more likely to become widely and rapidly adopted.

The efficiency gains achievable through dematerialization are illustrated by a study of the efforts of designers and manufacturers working to develop monoblocs. Development of monoblocs not only captured the imagination of designers and manufacturers but also delivered environmental benefits, through reduced material and energy requirements during production and transportation. The global success of the monobloc highlights the downside of dematerialization and injection moulding; cost savings, when passed on to consumers, drive sales. Increased sales of monoblocs and many other plastic items were reflected in higher overall environmental impacts, which quickly outstripped the savings achieved at the individual product level. However, the dematerialization dividends explored by pioneering designers working with injection moulding remain relevant today, as manufacturers strive to contain costs and reduce resource consumption while lessening environmental impacts.

While injection-moulded products including chairs were often relatively inexpensive, they lacked the 'aura' of fibreglass creations which proudly display their maker's craft skills.[20] For example, fibreglass chairs are dependent upon labour-intensive processes (sculpting moulds, lay-up, moulding, trimming, polishing). Injection-moulding's cost advantages swiftly marginalized fibreglass technology. Fibreglass chairs were cosigned to a high-end niche market, where the aura of hand-finished chairs is appreciated, despite their higher cost.

Although celluloid ultimately proved to be a poor ivory substitute to produce billiard balls, new fully synthetic plastics quickly gained acceptance by designers, manufacturers and consumers across many applications, including seating, following the end of the Second World War. Plastics led to a reduction in demand for scarce resources like ivory, helping to ensure the survival of the elephants and other species throughout the twentieth century. While plastics undoubtedly contributed to the elephants' reprieve last century, the long-term effects of extracting and consuming the fossil fuels required to make plastic, and the mess it has left us with, were about to become apparent.

2   Plastickoptimsimus and bust

At the end of the 1960s plastics reached their zenith in Europe and the United States. Jeffrey Meikle, author of *American Plastic: A cultural history* refers to this time as 'plastickoptimsimus'. When American won the Space Race it chose to commemorate the first moonwalk with a flag made from nylon (Figure 2.1). In fact, success in winning the Space Race owed much to plastics, as highlighted by Meikle, with the mission relying heavily on 'high-performance synthetics, on Mylar, Teflon and nylon, [as well as] heat-resistant composites and form-fitting foams' (J. Meikle, 1995, p. 216).

Figure 2.1 Edwin Aldrin next to the nylon flag used to commemorate Apollo 11 landing on the moon, 1969. Courtesy NASA.

Meanwhile, back on earth, in December 1967, in *The Graduate* Dustin Hoffman is advised, by his older lover's husband, that a lucrative future can be found in plastics. *Barbarella*, starring Jane Fonda in a futuristic erotic plasticized world, was released the following year as was Stanley Kubrick's *2001: A Space Odyssey*. At the same time, several of the world's most prominent art galleries and museums showcased exhibitions featuring plastic consumer goods, clearly demonstrating that the material, despite being neither rare nor precious, had acquired significant cultural value. In 1970, the Victoria and Albert Museum, London, presented an international exhibition of modern chairs dating between 1918 and 1970. The exhibition catalogue denotes that over a third of the 120 chairs (forty-eight) featured plastics (Victoria and Albert Museum & Whitechapel Art Gallery, 1971). In 1972, *Italy: The New Domestic Landscape*, the most ambitious exhibition yet presented at MoMA, featured 180 objects for household use and eleven environments commissioned for the event, all heavily featuring plastics (Ambasz, 1972). Plastickoptimsimus had peaked, or as Meikle claims, plastic reached its apogee (J. Meikle, 1995, p. 231).

1. Australia's plastickoptimsimus

Australia had to wait slightly longer than their contemporaries in Europe and the United States before reaching plastickoptimsimus. At the start of the 1970s, the trade press noted that 'plastics-as-plastics' were gaining popularity in the furniture industry especially among young designers and retailers targeting the youth market.[1] Traditional seating solutions, with their dull colours, arranged in formal fixed positions did not suit the times. Traditional materials and labour were becoming more expensive while plastics remained a cost-effective option (Shaw, n.d., p. 5). Seating requirements became more casual, adaptable and laid back. Modular seating components, which could be assembled and reassembled into a variety of combinations, became increasingly popular. Even bags full of plastic 'beans' became fashionable. In May 1973, Furniture Makers of Australia announced plans to introduce five new plastic ranges, but it was not a willing convert to the material, complaining to the trade press that it has been forced to expand its use of plastics as timber became more difficult to obtain and prices were going 'sky high' ('Fler Increase Their Use of Plastics', 1973, p. 35).

The mainstream press, having ignored plastic furniture for so long, suddenly embraced it as a symbol of the new, future-focused decade. An article in one of the leading homemaker titles headlined 'The plastic revolution' claimed:

For years, with a few exceptions, our wail has been 'Yuk, it's plastic'. Now the cry is changing to a joyful 'Wow, it's plastic!'

(Worsoe, 1974)

Both of the mainstream Australian homemaker titles regularly featured plastic furniture, even featuring it on three covers during the first year of the new decade (Figure 2.2).

Figure 2.2 Plastic furniture featured on the front cover of three issues of prominent homemaker titles in 1970. Courtesy *Australian House and Garden* October 1970, *Australian Home Beautiful* October & November 1970. Author's collection.

In 1941, Yarsley and Couzens predicted the life of a future 'plastic man'. Thirty years later Australians might well have been tempted to scoff at that vision of the future but, during the intervening years, plastics had quietly and successfully oozed into every aspect of daily life. Standing in their new home, Australia's plastic people might point out that their walls were not, as Yarsley and Couzens had predicted, made from plastic. Walls were, however, coated with acrylic paint and featured plastic light switches and power points. Those walls concealed a web of plastic plumbing and electrical wires insulated with plastic. Above their heads, plastic light fittings, and more acrylic paint beneath plastic waterproof membranes. Cabinets, tables and even the television were covered in laminates, with drawers and cupboards lined with sticky backed plastic.

The floor was either laminate, vinyl or varnished with polyurethane (PU). In the bathroom the bath, vanity, towel racks, shower curtain, toilet seat and fluffy cover were all made from pink or avocado pastel-shaded plastic with plastic soap trays surrounded by endless beauty and hygiene products, all bottled in plastic. In the kitchen, brightly coloured laminates and sticky backed plastic covered every surface and even the kitchen sink and washing-up bowl were plastic (Figure 2.3). Countless plastic canisters, Teflon-coated saucepans, Tupperware containers, cutting boards, melamine crockery, together with often long-forgotten 'labour saving' devices, unwanted plastic novelties and toys crammed the cupboards alongside the plastic Christmas tree. Dining chairs, easy chairs and sofas were not made from plastic but padded with PU foam and covered with vinyl or fabric laminated with thick clear plastic for protection. Under a plastic wall clock sat the plastic telephone on a laminated tabletop, resting on elegantly tapered tubular steel legs, powder coated with plastic. Rubbish, gathered in plastic bins, was taken outside and fed into plastic garbage cans discreetly positioned out of sight, adjacent to the plastic-lined swimming pool, past polyester and viscose laundry suspended by plastic pegs from a plastic clothesline. Plastic hoses watered plants in plastic pots. Weary gardeners sought respite on plastic outdoor furniture or retired to their beds to rest on plastic foam mattresses.

While plastic Australians probably did not think they would end their days in Yarsley and Couzens' plastic coffin, they most likely ended up in a medium-density fibreboard box veneered, with paper, painted with acrylics and varnished (with PU) to look like wood, with nylon or polyester lining padded with PU foam, all held together with polymer adhesives. Plastickoptimsimus had arrived

Figure 2.3 Acrylic sink advertisement. *Australian Home Beautiful*, June 1950. Author's collection.

in Australia but its reign was to be short-lived. Global events were about to impact the market.

2. Plastic love fades

In July 1969, Apollo 11 took off from the moon to return to earth, the rocket blast melting the plastic flag planted to commemorate the event (Tribbe, 2014, p. 9.). The year before, shortly after celebrating its decade-long run at Disneyland, Monsanto's completely plastic *House of the Future* was demolished. Young cinema-goers sneered cynically when Hoffman's character in *The Graduate* was advised to pursue a career in plastics (J. Meikle, 1995, p. 3). *NASA* announced, in 1970, the cancellation of the last three planned Apollo missions; public interest in the Space Race had fallen to the extent that few Americans could even name the first person to walk on the moon.[2] History professor Matthew Tribbe observed that sci-fi films switched their focus 'from utopian to dystopian' (*THX 1138, Logan's Run* and *Silent Running*) with terrifying portrayals of plastic futures (Tribbe, 2014, p. 209). Accusing someone of being plastic had become a common derogatory term implying, 'no personality, no depth, [and was] of mass production' (Leschen, 1978, p. 9). The plastic man envisioned by Yarsley and Couzens was no longer seen as a desirable reality, but increasingly regarded as a cautionary tale of how not to manage the impact of technology on the environment. The cult of technology and its ability to solve societal problems was increasingly interrogated.

Aware of its increasingly negative public image, the plastics industry introduced strategies to improve it. In 1971, the Plastics Institute of Australia joined forces with the Industrial Design Council of Australia to organize a touring exhibition *Design in Plastics*, specifically aimed at showcasing plastic products to 'counter any image of plastics as a "cheap and nasty" substitute for a better class of material' ('A Good Image through Good Design', 1971, p. 7). However, altering the opinions of consumers with first-hand experiences of poorly manufactured products proved more difficult than the industry imagined, particularly with the older cohort. A 1972 consumer market research study, evaluating attitudes toward plastic housewares, found that participants under thirty-five years of age were 'much more positive in their acceptance of plastic goods'. However, participants over this age displayed distrust, reflecting their early experiences with inferior plastic products (Porter, 1972, p. 19). The same study also found that plastic products were still generally considered to be of inferior quality. The era of plastickoptimsimus ended abruptly, the love affair with plastic was over.

The growth and prosperity of the post-war years stagnated, and social unrest swept across the United States and touched many allied countries, especially in Europe. OPEC members demonstrated their power and sent global energy markets into turmoil, raising concerns about dwindling fossil fuels (J. Meikle, 1995, p. 231). These exogenous events occurred as environmental and health concerns about plastics began to gain public attention. Additionally, the first seeds of doubt regarding the ability of the linear economy to deliver limitless growth emerged, ignited by Euston Mishan's 1967 book *The Costs of Economic Growth*. Where were rationalization, efficiency efforts and the privilege given to progress leading? The wholesale embrace of science and technology was increasingly questioned.

The oil crisis acted as a catalyst, causing significant disruption to the energy market, with impacts on the plastic industry compounded as the decade progressed. Initiated by the OPEC oil embargo, which started on 19 October 1973, the oil price quadrupled in just six months, taking the price of plastics with it, and disrupting the entire industry (Figure 2.4). Although the oil crisis was brief and ended in March 1974, the price of oil and plastics never returned to pre-embargo levels. The Iranian Revolution in 1979 further disrupted supplies with the Iran–Iraq War from 1980 to 1988 prolonging the crisis throughout the next decade. In addition, concerns about 'peak oil' and predictions that reserves could run out by the end of the century gathered momentum. Judy Attfield noted: 'the realisation dawned that some natural resources were non-renewable' (Attfield, 2000, p. 246).

In the short term, designers and manufacturers were forced to re-evaluate their use of plastic and many turned to traditional materials. This shift in consumer preferences made a significant impact on the furniture industry. In Finland, Asko gave up production of plastic furniture, leaving their star designer Eero Aarnio out of work (Aarnio & Savolainen, 2016). In Denmark, Fritz Hansen ceased production of Arne Jacobsen's Swan sofa (Hansen, 2017, p. 41). In France, Jean-Pierre Laporte abandoned his career as a plastic furniture designer and took up cabinet-making (Audibert, 2010, p. 121). While in the UK, Hille formed a partnership with

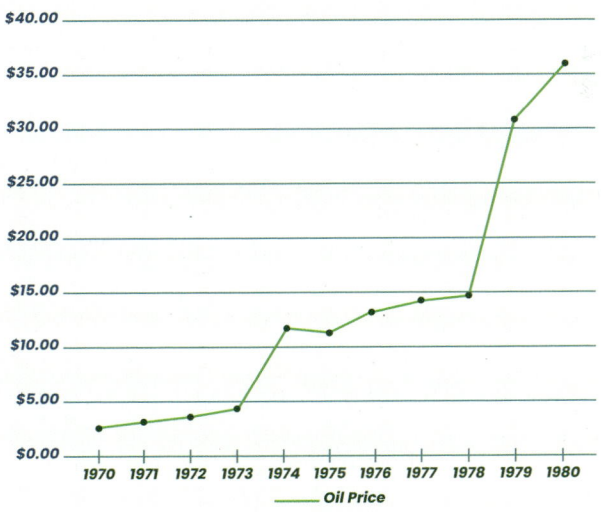

Figure 2.4 Impact of 1970 crises on oil prices. Adapted from World Bank commodity price data.

Tate & Lyle to investigate using bagasse, the waste from sugar cane, to produce furniture (Bony & Rakocevic, 2005, p. 80).

In Australia, Uniroyal found a way to work around problems caused by higher prices. They completely redesigned Grant and Mary Featherston's modular *Numero* series of PU foam chairs to reduce their reliance on the material by introducing a steel frame (Figure 2.5). The original *Numero* comprised just three components (seat, base and cover) and changing the design added considerable complexity and cost. The single-shot injection-moulded seat became a multi-part tubular steel frame with webbing manually assembled to support the minimum amount of foam needed to provide comfort. This expensive redesign provides a dramatic illustration of the impact of higher oil prices throughout the industry.

Figure 2.5 The *Numero IV*, Grant and Mary Featherston for Uniroyal, 1973. Courtesy *Australian House and Garden*, July 1974. Author's collection.

From its beginnings the synthetic plastics industry was associated with shoddily made goods. Despite efforts to counter this reputation, the fact remained that many plastic products were simply not fit for purpose. The industry maintained that it could not be held accountable where products have been poorly made or the wrong material had been used and complained that consumers blamed the material and not the manufacturer of the inferior product (Fleck, 1944, p. 99). But early plastics did frequently suffer from quality issues. The first semi-synthetic plastic, celluloid, can degrade irreversibly and can 'induce the breakdown of objects in its vicinity' (Dennis, 2020, p. 115). Cellulose nitrate had been used for film stock and could (and often did) explode (T. Fisher, 2013, p. 110). Early acrylic glass furniture became notorious for being easily scratched. Many early plastics were not UV resistant, colours bled or faded, others became brittle over time, while some (particularly foams) disintegrated completely. Remnants of these quality issues can still be found in mid-century furniture; for instance, PU foam padding under upholstery breaks down to dust after about forty or fifty years.

While the problems with early plastics took years or decades to reveal themselves, consumers soon realized that the flawless surface of plastic furniture was not as durable as expected. The utopian perfection of 1960s plastic furniture was expressed through its uniform glossy surface, which promised to resist time and show no signs of wear or patina as it aged. The public were assured that the flawless surface of plastics could be kept that way with minimal maintenance, requiring just a wipe from a damp cloth.[3] But the shiny, faultless surfaces quickly began to scratch and show signs of wear, losing their promise of perfection. 'The magic is broken', suggested Fisher, while Manzini claimed the surfaces 'degrade without dignity' (Manzini & Cau, 1989).

Glossy fibreglass and injection-moulded polyester were not the only plastics whose appearance deteriorated with use. Early plastic laminates such as melamine were particularly prone to showing signs of wear and tear. The decline in popularity of melamine tableware during the 1960s can be attributed to dissatisfaction with the staining and scratching that occurred with use of crockery produced in the 1950s (Akhurst, 2004, p. 8).

With traditional materials we accept and even celebrate signs of ageing – signifiers of past experiences. Every piece of wood has its own unique character, showcasing the story of its growth and location within a tree through attractive patterns of grain. Wooden furniture ages gracefully with use and exposure to light, and this constantly changing appearance is often considered appealing. Plastics, on the other hand, failed to live up to their promise of being resistant to wear and tear, with surfaces becoming easily scratched. The Dutch art critic Albert Plasschaert claimed:

> There is no real affection between a man and this type of [plastic] furniture; no memories will ever linger around them. They will never be like a familiar face in which we can perceive the reflection of our own personality. There is no intimacy.

(Taft, 2014, p. 48)

When abraded, glossy plastic surfaces trap dirt and stain easily leading to the plastic object eventually being discarded. Dirt can be trapped so effectively it cannot be removed by washing, quickly making the object less desirable. Unlike traditional materials, plastics cannot be repainted or repolished, with replacement the only option, further perpetuating the idea that plastic is synonymous with disposable, inferior quality goods.

The industry responded to quality issues by developing an ever-increasing range of additives to enhance desirable attributes such as strength while compensating for deficiencies, like fading. Reinforcing fibres, fillers, plasticizers, compatibilizers, surface modifiers, chain extenders, impact modifiers, coupling agents, scratch-and-mar resisters, colourants, ultraviolet stabilizers, flame retardants, peroxides, antioxidants and anti-stats were continually modified and optimized to improve the performance, appearance and stability of the material. Unfortunately, most of these additives are made from (often unregulated and frequently undisclosed) chemicals, at least 1,500 of which have been identified as chemicals of concern to public health (Simon et al., 2021, p. 44). Rachel Carson's enormously influential 1962 publication *Silent Spring* documented the environmental impacts of pesticides and received wide publicity, alerting consumers to the potential dangers of chemical additives more widely. The petrochemical industry attempted to discredit Carson's work but only succeeded in further raising public concern and anxiety toward the industry.[4]

The negative environmental impacts of plastics had gone largely ignored or unnoticed until the end of the 1960s. In 1968, Yarsley and Couzens published a completely revised version of their work examining plastics for a non-specialist audience, but no reference is made to any adverse environmental impacts (Yarsley & Couzens, 1968). This was typical of mainstream media industry coverage at this time, which made little reference to the environmental impacts caused by plastics. Attitudes toward the material were about to change dramatically. Buckminster Fuller published *Operating Manual for Spaceship Earth*, focusing attention on the planet's finite resources in 1969 (Fuller, 1969).

Paul Ehrlich drew attention to the contribution exponential population growth would have on environmental degradation (Ehrlich, 1968). In the same year, and for the first time, global environmental problems including pollution, resource loss and wetlands destruction were discussed by scientists from around the world at the UN's Biosphere Conference in Paris. On Christmas Eve 1968, Apollo 8 sent the Earthrise image back to earth where it featured on the front cover of the new counterculture magazine, *Whole Earth Catalog*. Hailed as the most influential environmental photograph ever taken, the image dramatically impacted public attitudes, alerting us to the finite resources available on this fragile planet.[5] The natural sphere and the human sphere were finally seen as co-dependent (Fallan, 2019, p. 21). Ecological awareness entered the public discourse.

In 1970, a week after the near-disastrous Apollo 13 mission returned, 20 million Americans observed the first *Earth Day*, and the Environmental Protection Agency was established in the United States. Barry Commoner published *The Closing Circle* the following year, which focused on sustainability and called for reduced use of plastics in both products and packaging. Public concern grew as landfills filled and plastics increasingly polluted the landscape. A growing unease with the excesses of consumerism began to gather momentum. The pristine luscious surfaces of plastics began to lose their appeal.

Design for the Real World: Human ecology and social change, 1971, by designer and educator Victor Papanek, drew attention to the environmental challenges created by the consumer society, focusing specifically on the role of the designer. Papanek referred to designers as a 'dangerous breed' in designing unnecessary things they are 'partially responsible for all types of pollution' (V. J. Papanek, 1971, p. 17). In 1972, an association of scientists and political leaders published *The Limits to Growth* for the Club of Rome. The report drew attention to the growing pressure on natural resources from human activities. Predictions emerged that the earth's limits would be reached in the following one hundred years if rates of population growth, resource depletion and pollution generation continued at the pace of the time (Meadows et al., 1972). Also in 1972, *Science* published the first article on ocean plastics reporting that 3,500 plastic particles had been counted per square kilometre in the western Sargasso Sea (Carpenter & Smith, 1972). The ability for bacteria to grow on the surface of plastics, the significance of the absence of plasticizers in the samples collected and the possibility that these had dispersed and entered the food chain were all discussed.

With the shine rubbing off the glossy surfaces of plastic, attention began to focus on the material that lay beneath. The suspected carcinogenicity of vinyl chloride monomer (VCM), an ingredient of polyvinyl chloride (PVC), raised concern about the potential health impacts of plastic in 1973. Those concerns were quickly substantiated when severe forms of liver cancer (hemangiosarcoma) were found in workers involved with the production of PVC (Westermann, 2013, p. 78). Reporting on the presence of phthalates (from PVC) in the blood of laboratory workers the *Washington Post* declared, 'we're all a little plastic' (Vogel, 2012, p. 48). By 1975, the industry was fighting a total ban on the use of PVC in food packaging in the United States. The following year the Australian Plastics Institute, keen to reassure consumers of the safety of PVC food packaging, wrote to the press: 'plastic food-packaging now manufactured from Australian PVC conforms to the maximum levels of VCM recommended by Federal Government authorities' (Eisner, 1976).

Health concerns started to emerge concerning other plastics. PU foam, by now widely used in household and office furnishings, cushions, pillows, mattresses, car and airline seats and as insulation, was shown to be highly flammable with the potential to release toxic fumes ('Plastics and You', 1976). This issue continued to plague the industry for many years and, in 1982, *Choice Magazine* reported that inflammable foam worsened the severity of house fires and released toxic gases, including hydrogen cyanide and carbon monoxide ('"Frightening" Risk in Foam Stuffing', 1982).

Early plastics often emitted strong odours, some of which consumers found alluring (Figure 2.6). Today the importance of smell is well understood by automobile manufacturers; Ford in Germany employs a 'Head of Smells' and Ford in Britain deploys an 'electronic nose' to check that cars are exuding the desired smell before they leave the production line (Piercy, 2002, p. 99). The new car aroma consists of a mixture of fumes from plastics, coatings and glues used in the assembly process. Maintaining the new car smell is so important to some owners that businesses sell aerosols specifically designed to recreate 'the musk of clean carpets, and the pure essence of clean plastic and rubber car parts and more that come together to make you relive feelings and sensations of excitement you thought long gone' (Chemical Guys, n.d.).

From the early 1970s, evidence began to emerge that inhaling these aromas (particularly those emitted by early plastics) might well cause detrimental health impacts. In 1972, Australian scientists began investigations into possible health dangers from inhalation of these fumes, including the 'new car smell' ('Plastics Fumes "Hazard"', 1972). In the case of thermoplastics like PVC, the odour is produced as the volatile elements of the material evaporate, a sign of

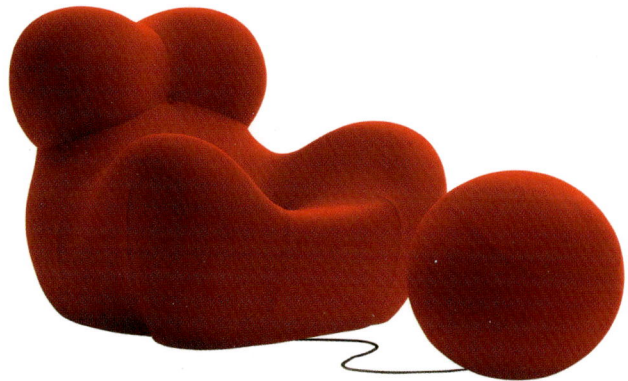

Figure 2.6 *Up* series, Gaetano Pesce, 1969. The original version arrived flat-packed and began to self-inflate once the airtight packaging was pierced. Production ceased in 1973, when it was realized that Freon, the leavening ingredient added to the polyurethane and emitting a futuristic sweet chloroform smell, was harmful to the ozone layer. Production recommenced in 2000 with a Freon-free version (pictured) which, sadly, no longer arrives vacuum packed. Courtesy B&B Italia, www.bebitalia.com.

degradation leading to decreased plasticity. Other early plastics could produce unsavoury odours. In 1964, author (and long-time critic of plastics) Norman Mailer characterized the United States as 'a sick nation, we're sick to the edge of vomit and so we build our lives with materials that smell like vomit, polyethylene and Bakelite and fibreglass and styrene' (Gabrys et al., 2013, p. 111).

In a very short space of time, the public image of plastics changed. Quality, health and environmental concerns shifted consumer attitudes toward plastic. Glossy shiny surfaces lost their lustre and were no longer admired for their perfection but seen as 'potentially oozing poisons' (T. Fisher, 2013, p. 300). Plastic-as-plastic furniture went out of fashion and the industry lost interest in developing plastic chairs. Following the burst of innovation during Space Age there was a dearth of new designs for plastic chairs until the end of the twentieth century. The price differential between plastics and traditional materials had been eroded and threated to impact demand for plastic.

3. Crisis? What crisis?

Despite the significant impact of these crises the effect on the overall demand for plastic was negligible. The local trade press noted that demand had been mostly unaffected by the price increases caused by the energy crisis. Oil, and therefore plastics, were more expensive, but still competitively priced when compared with alternative materials ('The Prospects for Plastics', 1974, p. 3).

In 1975, the production of plastics consumed only 2 per cent of the oil produced and the industry complained that despite its negligible impact on energy markets, it was being treated as the 'whipping boy' for the crisis (Pickering, 1975, p. 6). The industry need not have been concerned. While demand was dented in the years immediately following the energy crises, growth quickly resumed and continued unabated until the global financial crisis (GFC), of 2007/2008 which also had only a limited impact on demand, as can been seen in the graph in Figure 2.7.

While overall demand for plastic remained buoyant, interest in furniture celebrating plastic-as-plastics dwindled. Glossy shell chairs, which had featured prominently in interior shots and advertisement in the homemaker magazines during the first three years of the decade, virtually disappeared.[6] By 1979 even

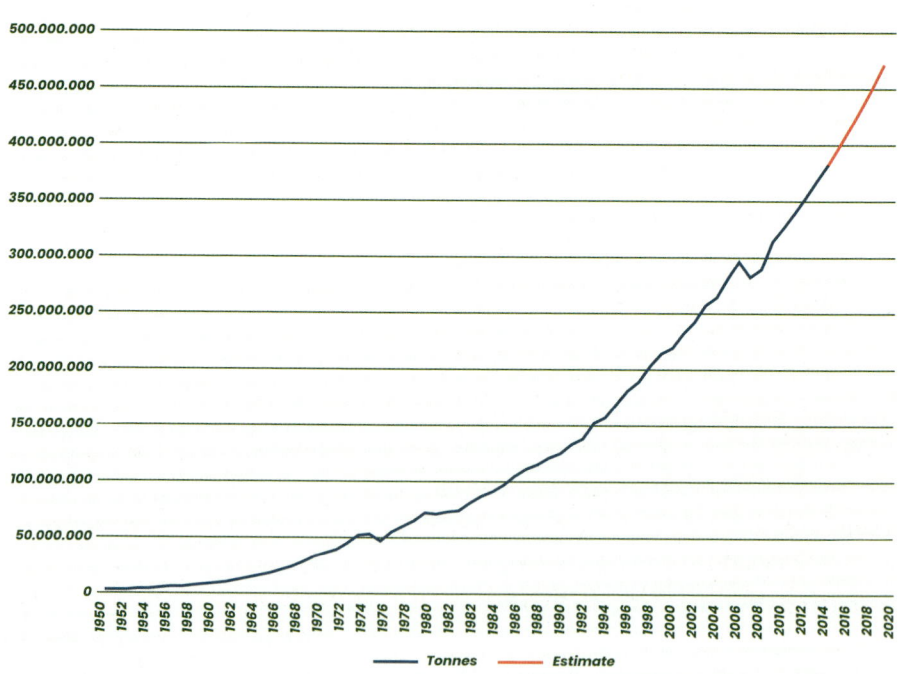

Figure 2.7 Graph showing plastic production 1950–2020. Adapted from Geyer et al. (2017b).

industry players were admitting defeat with a local General Manager exclaiming, 'plastics has had it as a fashion story. It has found its place as a purely structural material for the garden and office furniture' (Martin, 1979). This reflected a global trend with Tribbe reporting: 'Space Age lifestyle became a future that few Americans any longer embraced' (Tribbe, 2014, p. 212). He reports that the 'natural' look made a comeback, with rustic earth tones like avocado and ochre replacing bright bold colours as the preferred aesthetic. The sleek Space Age style was replaced by a handcrafted look with an emphasis on traditional materials (Tribbe, 2014, p. 212). Australia followed a similar trend, with the bright colours associated with the Space Age quickly giving way to more muted tones. Anne Watson noted that 'exuberant colour also fell victim to the conservatism engendered by the changed economic conditions' (A. Watson, 2002, p. 13). Penny Sparke suggests the 'ecological outcry of the 1970s' led to plastics being considered as inferior to natural materials (Sparke et al., 1993, p. 54). In the UK, a revival of interest in Victorian and Art Deco interiors also came at the expense of plastics, according to curator Mark Suggitt (Lambert, 2021, p. 170).

Despite these changes in consumer preferences, the plastics industry retained its position inside the home by stealth. Plastic-as-plastic had become unfashionable, but upholstered modular furniture continued to sell well for the remainder of the decade and into the 1980s. Ironically, the modular lounge units were typically made from PU foam and, as the upholstery needed to stretch in two dimensions, the fabric was invariably plastic based. Modular bookcases and shelving remained popular but local suppliers began to emulate wood finishes to accommodate the change in tastes. Other apparently 'wooden' furniture was increasingly likely to similarly be made from veneered composite boards held together with polymer adhesives, varnished with PU with plastic glides to protect wooden floors (polished with PU). Inside the home, plastics maintained a presence, often disguised, or mixed with other materials to make them less obvious. While plastic kettles fell from favour, metal models featured plastic handles. The industry had simply revisited the strategy that succeeded in getting the material accepted in the first place – plastics were integrated or camouflaged so that users barely noticed them. Industrial designer Sebastian Conran even invented his own term 'deplastification' to describe this, then fashionable, action of making things feel and appear less 'plasticky' (Lambert, 2021, p. 104). In short, plastics became invisible (T. Fisher, 2015, p. 120).

4. Plastic revival

In June 1992, 152 world leaders signed conventions on biological diversity, deser-tification, a framework on climate change and principles for sustainable forestry at the Rio de Janeiro *Earth Summit*. The report accused industrialized countries of pursuing a pattern of consumption and production which aggravated poverty and wealth imbalances.[7] The ability of linear economies, dependent on high levels of production, consumption and waste, to deliver endless growth became increasingly questioned. The role and responsibilities of professional designers in supporting the system came under increased scrutiny. Papanek's *The Green Imperative* (1995) claimed that the 'ethical' role of design involves considering the environmen-tal consequences of products, and he urged industrial design professionals and end-users to recognize their ecological responsibilities (V. J. Papanek, 1995, p. 2). He accused designers of being complicit with marketing professionals in produc-ing and making unnecessary and wasteful objects (V. J. Papanek, 1995, p. 17).

In the decade leading up to these events the plastics industry undertook intensive public relations initiatives and spent millions of dollars on supporting advertising campaigns promoting the effectiveness of recycling and reinforcing the belief that plastic could be continuously and economically recycled at scale. By March 1995, a trade publication declared the 'plastics recycling war is over', with the public accepting recycling as a bona fide solution to the plastic waste issue. An industry presentation claimed that plastics were now seen as comparable to competitive materials (e.g. paper, glass, and aluminium) among the general public (Centre for Climate Integrity, 2024, p. 21).

These changing attitudes were reflected in the furniture industry when, after an absence of nearly two decades, a plethora of new designs for plastic chairs appeared on the market toward the end of the twentieth century. Kartell collab-orated with Philippe Starck (b.1949) to develop a translucent monobloc using polycarbonate (PC) (Figure 2.8). *La Marie* made its debut at the *Salone del Mobile* (Milan, 1999), with chairs displayed in a circle, withstanding heavy blows from a rotating industrial robot. Designed to demonstrate both the strength and flex-ibility that belie the design's petite appearance, the publicity stunt successfully ignited interest in PC furniture from both critics and retailers alike. Weighting just 3.5 kilograms and boasting the transparency of glass, *La Marie* embod-ies Philippe Starck's commitment to dematerialization (Wingfield, 2018). His success in improving both material and energy efficiency was the result of fifty

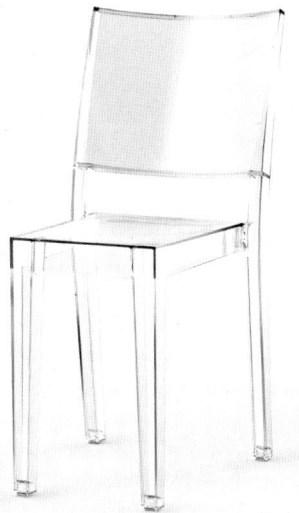

Figure 2.8 *La Marie*, Philippe Starck for Kartell, 1998. Courtesy © Kartell.

years of industry experimentation with dematerialization combined with sixty years of research and development in the plastic furniture industry backed by the financial resources of plastic specialist manufacturer Kartell (Morello & Ferrieri, 1988; Storace et al., 2012).

During extended production experimentation, Kartell and Starck discovered PC was better suited to moulding straight-edged designs. Thus, the final form of *La Marie* marked a return to geometric forms favoured by early modernists such as Breuer and Le Corbusier (Fiell et al., 1993, p. 12). This style had remained virtually absent during the first sixty years of plastic chairs, with designers having favoured exploration of organic curves. Transparent furniture quickly became a popular choice for furnishing apartments and small spaces, as it de-clutters while blending into any style. Interior designer Sormeh Rienne summarized the broad appeal: 'whether you're a minimalist or a maximalist, this chair works' (Lauren, 2019). The design has also proved popular with museums and art galleries around the world, as the chairs recede and do not compete with items on display.

La Marie marked a turning point for the industry and transparency has since become a major feature of Kartell's catalogue, with numerous designs added. The most popular of these, *Louis Ghost* (2002), has sold over 3 million copies and has also been widely copied (Figure 2.9). Sixty years after the first acrylic

Figure 2.9 *Louis Ghost*, Philippe Starck for Kartell, 2002. Courtesy © Kartell.

glass chairs, transparent furniture finally became widely available and accepted by the mainstream market. Competitive pricing from Kartell contributed to the commercial success of the transparent series.[8] Transparent furniture also took Starck closer to his stated ideal of 'democratizing' design – improving quality while striving to make it accessible to the greatest number of people, at affordable prices (Wingfield, 2018).

Rashid also used PC to develop his *Oh Chair* (1999), again enjoying enormous commercial success with over 6 million copies of the low-priced design sold (Figure 2.10). Against a background of health and environmental concerns, the success of those PC designs is yet another example of plastic's paradox. PC lacks a unique resin identification code (RIC) and is classified as 'other' (RIC 7), signalling that it is unlikely to be recycled. Despite this, the chairs continued to sell by the millions.

Also in 1999, Jasper Morrison (b.1959) fêted the introduction of the next step-change injection-moulding technology to the furniture industry by naming his design after it (Figure 2.11). Air injection technology made it possible to create highly resistant, super-lightweight products and eliminated the need for structural ribs to provide strength. The owner of manufacturer Magis demonstrated the capabilities of the technology by showing Morrison a section of tube made using

Figure 2.10 *Oh Chair*, Karim Rashid for Umbra, 1999. Courtesy Karim Rashid Inc for Umbra.

Figure 2.11 *Air Chair*, Jasper Morrison for Magis, 1999. Courtesy © Jasper Morrison.

the air injection process. Realizing the lightweighting potential, Morrison experimented with the new technology leading to the development of the *Air Chair*.

While the technology had been used to make components by the automobile industry, air injection was relatively new to the furniture industry.[9] Nitrogen gas is pumped into the mould to create hollows in the still-molten plastic, minimizing the PP required (*Air Chair* weighs 4.4 kg). The pressured gas also ensures the plastic does not shrink away from the mould's surface, preventing warping when complex and challenging forms are being created. Gas helps distribute the pressure evenly, eliminating the stress and flow lines that can occur during a standard injection process.

Monoblocs typically provide adequate strength in their legs through folding or wrapping a sheet of material. Air injection enabled new approaches, freed from the need for uniform thickness as John Tree, senior designer at Japer Morrison's office, explained to me:

> It was plastic becoming flexible enough to do very simple designs that allowed the exploration of the kind of typology of the object, rather than have the material dictate its form … before that time, no one would consider this possible in plastic because of the size of the sections and the thickness and the use of gas really was what opened those possibilities up.

Air injection quickly became widely adopted by the furniture industry, with manufacturers attracted by the potential for reduced unit costs. While upfront tooling costs are relatively expensive, marginal costs are low, although a longer cycle time is required on the moulding machine compared with traditional injection processes. Material and transportation costs are also lowered due to the lightweighting effects of the process.

Also, in 1999, Ron Arad developed the *Tom Vac* chair for Vitra. The previous year had seen the release of Riccardo Blumer's *Laleggera* for Alias, Mario Bellini's *Bellini Chair* for Heller and a plastic version of Marc Newson's *Orgone Chair* from Artificial. Environmental concerns were suppressed by the weight of industry reassurances that recycling would produce a technological fix to the waste problem. Plastics were back in fashion. The furniture industry quickly adapted to these shifting tastes, with leading designers commissioned by major manufacturers to continually develop new designs throughout the first decades of the twenty-first century.

The next major technological advances to be embraced by the industry emerged from the digital realm. London-based design duo Barber Osgerby (Edward Barber and Jay Osgerby) developed the *Tip Ton Chair* specifically for the education market (Figure 2.12). The chair can tip between two positions simply by the sitter shifting their weight. The forward-leaning sitting position is specifically designed to straighten the pelvis and spine. The skid base abolished the expensive levers, knobs and mechanisms required by many adjustable chairs designed for workstation use. The *Tip Ton Chair* (2011) is among the first chairs developed using flow simulation software. Edward Barber (b.1969) explained the advantage of incorporating the digital technology and how the software enhances the efficiency of the injection-moulding process:

> Vitra had recently acquired new software that enabled them to [conduct] very accurate mould flow simulations. That enabled them to make the sections really quite thin, as small as possible. We could simulate what the outcome was going to be before they cut the tool and this was quite new for them at this point ... this was the first time they [Vitra] had done a whole plastic chair using this simulation. In the past, they always had to overcompensate because you invest so much money in a tool you can't afford to have a chair come out and it's not strong enough.

Figure 2.12 *Tip Ton Chair*, Barber Osgerby for Vitra, 2011. Courtesy © Vitra.

The design of Colombo's *Universale* (1969) had to be modified during pre-production, as the hole in the back of the chair interfered with the flow of material in the mould (Morello & Ferrieri, 1988). Flow simulation software enables similar issues to be identified before the design advances to the tooling stage, potentially delivering significant cost savings. More efficient use of materials and the ability to develop ideas reliant on thinner components contribute to the rapid adoption of flow simulation software by the furniture and consumer goods industries.

5. Summary

From the late 1960s, the negative ecological and health impacts caused by both the creation and disposal of plastics became increasingly apparent. The oil crisis and disappointing experiences with early plastic products of inferior quality, combined with a rising ecological undercurrent, prompted a re-evaluation of the cultural perception of plastics. Designers, manufacturers and consumers all reassessed their relationship with plastics. These shifting attitudes were reflected in the furniture industry, with relatively few significant designs of plastic chair launched between the mid-1970s and the late 1990s. However, overall demand for plastics continued to grow as manufacturers incorporated (often disguised) plastics into an ever-increasing variety of household products. This trend for 'deplastification' declined toward the end of the twentieth century, with products obviously made from plastic, including chairs, returning to favour following industry efforts promoting the myth of continuous recycling. Despite mounting environmental and health concerns, the short-term compelling advantages offered by plastics prevailed.

As plastics returned to favour, designers and manufacturers working in the furniture industry revisited the quest for dematerialization, leading to the development of several iconic chair designs with significantly reduced resource demands. Although primarily driven by profit, this focus on dematerialization coincided with emerging contemporary sustainability initiatives encouraging innovation and efficiency as a strategy to address environmental problems. Paul Burall's *Product Development and the Environment* reflected the thinking of the time, asserting that environmentally responsible product design was in line with managerial objectives, as both sought the most efficient use of resources (Burall, 1996). Designers, manufacturers and consumers were all becoming increasingly aware of the true value of doing more with less. While the focus of sustainability discourse has since shifted from the product level to the system level (as discussed in Chapter 6), the insights gleaned from decades of experimentation with dematerialization remain pertinent and essential today, as the world endeavours to find solutions to the environmental emergency.

By the end of the twentieth century, fossil-based plastic technology had matured, and dematerialization was nearing its limits. While advances such as flow simulation software continued to improve efficiency and reduce environmental impacts, they were iterative, and the era of step-change improvements, such as those brought by air injection moulding, had come to an end. Dematerialization efforts continue to this day, with promises of energy efficiency savings from the latest injection-moulding machines, modifications to hot runners and screw designs focused at reducing production costs by cutting resource consumption. But dematerialization has its limits: plastic chairs, for example, simply cannot get much smaller, lighter, thinner or even less visible. As Starck observed, 'you cannot dematerialize a chair completely, because you must continue to sit on it' (Blum, 2008). The promise of infinite progress is eventually trumped by the law of diminishing returns.

As the industry entered the twenty-first century, it became increasingly apparent the environmental benefits achieved through dematerialization were continually outstripped as demand for goods and services grew, driven by an increasingly numerous and affluent global population (Escobar, 2018, p. 33). As issues critical to humanity, including the exhaustion of fossil fuel resources, GHG emissions and climate change, became of increasing concern, the need for a new approach became ever more apparent. The need to take action gained consensus (IPCC – Climate Change 2001, 2001). Attitudes towards plastics have shifted again as the full health and environmental impacts caused by their creation and disposal are better understood and the limited impact of mechanical recycling has been revealed.

Plastics are now increasingly seen as an emblem for everything that is wrong with mass consumption, yet the material remains an indispensable part of our daily lives (Westermann, 2013). This plastic paradox represents a 'wicked problem' that has remained unresolved for over a century. During that time the negative impacts of plastics have been slowly accumulating and now, faced with our own threat of catastrophe, demand urgent attention. The next chapter examines the environmental impact of the petrochemical industry's current growth plans, discusses alternative future scenarios and emphasizes the need for a rapid increase in the use of alternative materials to replace virgin plastics.

3 Just say no to virgin fossil plastics

Of the total 8.3 billion tons of virgin plastic produced by the end of 2020, the majority, 59 per cent (4.9 billion tons) ended up in landfills, where it will remain for decades, if not centuries to come, or littering our environment (Geyer et al., 2017a, p. 1). This chapter examines the petrochemical industry's plans to continually grow the market for virgin plastics through to 2050. With plastic production projected to increase every year, the GHG emissions from the industry cannot be ignored any longer, especially as these plans are not compatible with the UN's GHG emission reduction targets. Due to the increasing volume of GHGs and toxic chemicals released during the production of plastics, it is essential to find ways to lessen the damage caused by these materials if we are to solve the climate crisis (Storrow, 2020). Plastics and climate change are inextricably linked; solving the plastics problem is crucial to addressing the climate emergency.

To address this need, an alternative scenario is proposed, focusing on the development and use of recycled plastics and bioplastics, collectedly referred to as renewable carbon-based plastics (renewable plastics). This analysis demonstrates that widespread adoption of renewable plastics is crucial to achieving real progress toward a more sustainable future. Renewable plastics are currently being used primarily to replace packaging, previously made from virgin plastics. But packaging consumes only about a third of the plastic produced. Development of a more sustainable plastics industry is reliant on consumer acceptance of products made using renewable plastics. Designers and manufacturers must be prepared to experiment with renewable plastics to develop more products with lower environmental impacts that appeal to consumers.

1. The plastic paradox

Industry proponents highlight the substantial economic and environmental benefits offered by plastics over traditional materials. For example, it is estimated that replacing PET with glass to bottle soft drinks would increase transportation costs by to five times and consume 40 per cent more energy.[1] Boeing claims their Dreamliner consumes 20 per cent less fuel than their previous generation of jets, mainly achieved by replacing 1,500 riveted aluminium sheets with lighter carbon fibre composites (Boeing, 2014).

Plastics also feature unique benefits that make them irreplaceable for many applications. Food can be kept hygienic and fresher for longer, reducing waste by up to half while delivering household savings, and reducing methane emissions from landfill.[2] Our health system depends on plastics, from PPE designed for our safety, to the tubes and machines that keep us alive in a time of need.[3] As we have seen in the previous chapter, plastics are now an indispensable part of our daily lives. There is little to be gained, and often much more to lose, by reducing or banning the use of virgin plastics without precisely specifying the materials that should be used to replace them for any given application.

Designers participating in research for this book often emphasized that when used appropriately, plastics can be the most environmentally sustainable material choice. Industrial designer and academic Manuel Garcia reported:

> Plastic is a material that can be revitalised … We have a huge effort to get people to understand plastic is actually a viable material and can contribute to our sustainability level in our planet.

Well-designed, well-made plastic products, including chairs, can last decades. Even at their end-of-life plastic products often have the potential to be recycled.

a. The real costs of plastic

However, the downstream impacts caused by the inappropriate disposal of plastic, particularly single-use plastics (SUPs), are increasingly the cause of public concern. The past decade has seen intensified focus on the problem of ocean plastics. The ever-growing quantities of plastic waste circulating in ocean gyres have increased awareness of the problem (see Lebreton et al., 2018). A report from the Ellen MacArthur Foundation forecast that if current trends continue, the weight of plastic in the ocean will surpass the total weight of fish by 2050 (Ellen MacArthur Foundation, 2017). This alarming prediction received widespread media coverage and stirred consumer backlash against plastics. The phenomenon of 'plastiphobia' has been further fuelled (particularly in Europe but also in Australia) by David Attenborough's *Blue Planet II* programme.[4] The final episode of the series devoted six minutes to the impact of plastic on sea life, showing a turtle hopelessly tangled in plastic netting, and an albatross killed by ingesting plastic shards. This segment generated the biggest reaction of any in the entire series, according to the BBC (Buranyi, 2018).

Ocean plastics are the subject of increased scrutiny from both the academic community and legislators (Nielsen et al., 2020). While the amount of research and policy attention given to ocean plastics is most welcome, it risks detracting from other critical issues such as climate change and overfishing, which are considered by some to be greater environmental threats (Stafford & Jones, 2019, p. 187). While the media and public are distracted by ocean plastics, governments and corporations can earn goodwill and appear proactive by addressing a symptom of the plastics issue while foregoing action in other areas of urgent environmental need (Stafford & Jones, 2019, p. 187). This singular focus on ocean plastics distracts attention from the upstream links between plastic production and climate change. Instead, the focus remains on downstream impacts, a relatively small symptom of a much larger plastics problem. A review of 180 scientific articles on plastics drawn from the fields of environmental science and environmental studies found a similar pre-occupation with a focus on marine pollution, leading the authors to a similar conclusion:

> If the debate continues to constrain itself to marine plastic pollution, plastic producers may feel free to ramp up the total supply of plastics entering the system while attention remains fixed on straws, cup lids, and sea turtles.
>
> (Nielsen et al., 2020, p. 13)

Ocean plastics are merely a symptom of a much bigger waste problem – 'the highly visible tip of the iceberg', as one study described them (Hundertmark et al., 2019, p. 9). Although it should be highlighted that most ocean plastics are not floating round on the ocean surface but are in fact microplastics or nanoplastics which are mainly invisible. Focusing on ocean plastics also risks ignoring what happens to the remaining 95 per cent of discarded plastic. Between eight and 12 million tonnes of plastic are estimated to be entering waterways every year

(Ellen MacArthur Foundation, 2017; Jambeck et al., 2015). This represents only about 5 per cent of the 200 million tonnes of plastic waste estimated to have been generated during 2020, most of which ends either up in landfills, is incinerated or litters our environment.

Growing awareness of these downstream impacts has led to 'A War on Plastics' being declared led by the BBC (kick-started by a television programme of the same name) and, to a lesser extent, the ABC in Australia. Consumer outrage has been focused on SUPs, leading to an increasing number of jurisdictions banning plastic bags and/or other SUPs. By the middle of 2020, seventy-seven countries had banned single-use plastic bags and thirty-two charge a fee (or tax) to discourage their use. Plastic Free July, encouraging consumers to avoid SUPs for a month, enjoys growing support from the global community with organizers estimating 140 million people across 190 countries participated in 2021 (Plastic Free July, 2021, p. 6).

All this focus on end-of-pipe littering impacts risks drawing our attention from the unsustainable trajectory of our consumption of plastic. We are failing to tackle the underlying issue, our dependence on fossil fuels. Too little academic attention has been given to investigate 'how the material properties of plastic are inextricably bound up with our dominant systems of production and consumption' (Nielsen et al., 2020, p. 13). An analysis of more than 180 papers found that academics engaged with the natural sciences and engineering 'are heavily represented in the upstream, production-oriented literature, dealing mainly with technical concerns', while research by social scientists focuses downstream on consumption and waste. By ignoring upstream impacts, societies based on disposability and overconsumption are maintained. The paper concludes by suggesting that 'if social scientists were encouraged to look further upstream, we might also note an increasing politicization of plastic production in the future' (Nielsen et al., 2020, p. 13).

Nielsen's observations on the self-imposed limits of academic enquiry also apply to the relatively new field of transition studies. Energy is often the primary focus of studies designed to map a course to a more sustainable society. The transport regime has also received much attention, with electric vehicles and biofuels evaluated for their potential as offering more sustainable solutions. However, focusing on energy and transport alone is insufficient to deliver the savings needed to reach global emissions targets. Chemical production is set to become the single largest driver of growth in global oil consumption by 2030 (Nova Institute, 2020b, p. 7). While the energy and transport regimes already have access to technologies that can enable them to decarbonize, the petrochemical industry remains dependent on consuming ever-increasing quantities of fossil fuels. It is not possible to decarbonize organic chemistry; nearly all plastics are made from carbon, a problem that demands our urgent attention.

The plastics industry actively supports efforts to maintain focus on the downstream impacts of plastic by supporting initiatives such as Keep America/ Australia Beautiful to point the finger of blame at the consumer as the source of litter. This diverts attention away from the plastic producers and their clients, the true source of the plastics pollution crisis. A Changing Markets Foundation report from 2020 presents case studies from fifteen countries across five continents to illustrate the activities of the ten largest plastic polluters and their attempts to obfuscate the plastics crisis (Changing Markets Foundation, 2020). The report also details the plastics industry's ongoing efforts to block anti-plastic legislation. The success of these efforts is reflected in current industry forecasts which anticipate continued growth in demand for plastics. Revealing the defuturing of the current regime is the first step to confronting change.[5]

2. Future scenarios

Industry projections expect demand for plastics to more than triple to 1.5 billion tonnes by 2050 (Figure 3.1) (Ellen MacArthur Foundation, 2017, p. 24). The petrochemical industry has enjoyed a compound annual growth rate (CAGR) of around 8.4 per cent for plastics so far this century (Geyer et al., 2017a, p. 1). A CAGR of at least 3.5 per cent is forecast to continue until 2050, despite the short-term impact of Covid-19 or the activities of governments and not-for-profits to reduce plastic consumption. Demand is forecast to continue to grow (American Chemistry Council, 2019, p. 10), driven by an increasingly affluent, populous and urbanized population.[6] To meet this demand, GHG emissions from plastic production will increase from 4 per cent to 15 per cent of the available global annual carbon budget by 2050, making it virtually impossible to reach global emissions-reduction targets (Ellen MacArthur Foundation, 2016, p. 24).

Increased production of plastics will drive nearly half (45 per cent) the forecast growth in demand for oil during the next three decades (International Energy Agency, 2018). The relatively small plastics market is then crucial to maintain the global petrochemical industry's growth. But oil will not be the only feedstock used to meet burgeoning demand for plastics. Gas has become plentiful due to

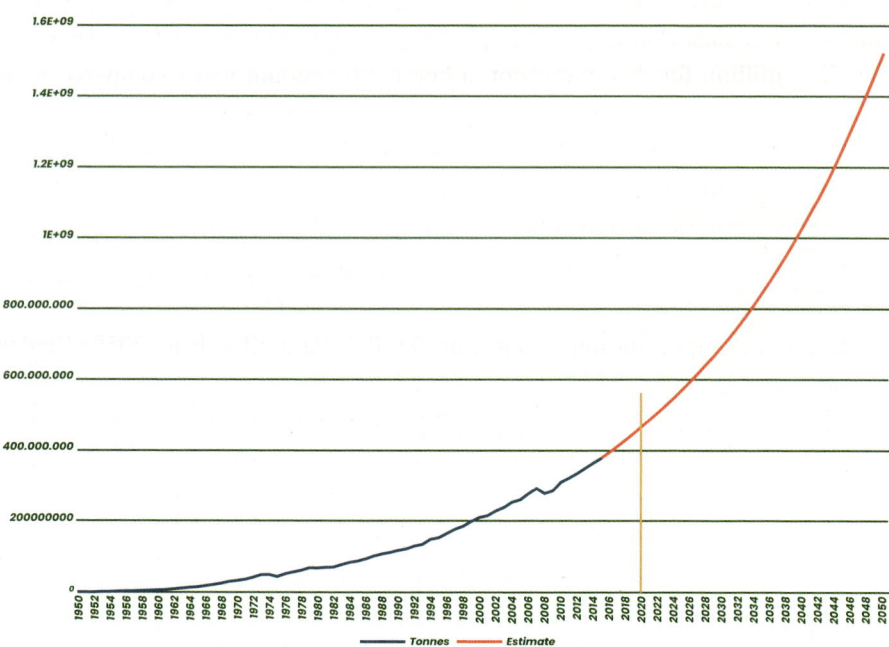

Figure 3.1 Global plastics production and forecast to 2050. Adapted from Ellen MacArthur Foundation (2016) and Geyer et al. (2017b).

dramatically increased fracking activities across the United States with ethane production exceeding 2.7 million barrels per day in April 2023, and the petrochemical industry is adapting to exploit this opportunity. To make plastics from gas requires investment in cracking plants to convert ethane released during shale oil extraction into ethylene. After being separated from natural gas liquids ethane is fed into a cracker where it is exposed to high temperature (850°C) and pressure, thereby splitting or 'cracking' the ethane molecule into ethylene.

The petrochemical industry has committed investment of between $200 and $300 billion in more than twenty new ethane cracking plants across the United States (Bruggers, 2019, n.p.; Center for International Environmental Law, 2019, p. 8; Food & Water Watch, 2019; Vidal, 2020, n.p.). One mid-sized plant, costing $6 billion, is expected to produce enough GHG emissions to effectively wipe out all the reductions in carbon dioxide planned by the nearby city of Pittsburgh in their efforts to combat climate change by 2030 (Bruggers, 2019). This Shell complex was expected to become the second biggest emitter of volatile organic chemicals (VOCs) in the state (Frazier, 2022a). However, in just one month, during 2022, the Shell facility emitted 512 tons of VOCs, against the permitted amount of 516 tons over a 12-month period (Clean Air Council, 2022). Shell was fined $10 million for this overshoot, a negligible amount when compared with the $1.6 billion in tax credits granted to the company by the state of Pennsylvania to incentivize development. Quite remarkably, following years of construction, Shell chose to announce the commencement of operations just days before the United Nations Climate Change Conference, COP27 (Frazier, 2022b). This one complex represents a small part of the expected $1.4 trillion to be invested by oil and gas companies between 2020 and 2024 to finance new extraction projects, effectively locking in our dependence on fossil fuels until at least 2050 (Center for International Environmental Law, 2019, p. 2).

Virgin plastics represent a valuable niche market for the energy industry but they are essentially a biproduct from the extraction of fossil fuels. In the case of oil, a small fraction of each barrel is effectively allocated to making chemicals, including plastics, supporting the economic advantage of maintaining oil in the energy mix. Energy analyst Simon Bennett highlights the fact there are only about 700 oil refineries worldwide, controlling both the fuel and chemical markets. The owners of these facilities have entrenched interests, resisting a shift to renewable feedstocks (Bennett, 2012a, p. 351).

Gas has emerged as an increasingly common feedstock for plastic since the Bennett paper was published in 2012. The principle remains unchanged, plastics represent a profitable additional niche market to support fracking operations. A small number of multi-nationals dominate the fossil fuel markets and many of them have interests in (or control) petrochemical organizations, increasing pressure to maintain the status quo. While plastics are produced all over the world, just ten organizations dominate the market with only about one hundred significant players (emphasizing the need for a global solution). It is in the interest of these organizations and their shareholders to maintain business as usual.

Reduced production costs achieved by switching to gas will allow the price of virgin plastic to be cut, shoring up demand while effectively pricing recyclates out of the market. During the second half of 2019, European prices for virgin PET collapsed and were lower than for recycled PET (rPET) (International Energy Agency, 2020, p. 87). When the cost of virgin plastics is low it becomes difficult to mount an economic argument to justify recycling. In 1991, there were twelve PE resin recovery facilities operating in the United States, by April 2015 none were left (Spalding & Chatterjee, 2017, p. 1248). During 2022, bales of scrap

PET, high density polyethylene (HDPE) and PP all continued to trade well below their four-year average prices (Recycling Markets, 2022). In 2023, prolonged lower prices for recyclates together with elevated energy prices, driven by the ongoing war in Europe and OPEC output restrictions, were applying further pressure to margins.

While forecasts can be useful as an alarm to warn us when the prevailing trend will lead us to an undesired future state, they are not effective as a planning tool and ineffective in guiding our actions (Höjer & Mattsson, 2000, p. 428). Despite the efforts of David Attenborough, the Ellen MacArthur Foundation and countless other organizations and community groups to call for action, demand for plastics continues to grow unabated.

To avoid the potentially devastating scenario being mapped out by the petro-chemical industry an actionable plan for radical change is necessary, starting with a rejection of virgin plastics. We must start saying no to virgin plastics. Transition management literature places significant emphasis on the important role that 'guiding visions' play in directing efforts and interventions toward more sustainable solutions (Smith et al., 2005). Individual actors are not capable of changing the entire system, nor can they be expected to identify the desired direction of change. It is difficult and often prohibitively expensive for an individual actor to evaluate and compare the environmental impact of decisions where multiple options exist.

Providing a meta-vision can guide actors, be they designers or manufacturers, to deal with the challenge of creating fundamental change (Schot & Geels, 2008, p. 542). Visions map paths toward plausible future alternatives by provoking debate and providing guidance on system innovations, highlighting the technical, institutional and behavioural problems that need to be resolved. Ambitious strategic visions covering the entire value chain can spur innovation and help secure funding. Once a vision is established, organizations can revise policy settings and identify the actor-network participation required to execute the desired change, allowing targets and goals to be defined and progress monitored. An overarching vision provides direction for participants toward the same goals and helps them secure the external resources and support required to achieve them.

The UN has established two high-level sustainability goals with potential impacts on the plastics industry. Sustainable Development Goal 12.5 targets substantially reducing waste generation through prevention, reduction, recycling and reuse by 2030. While goal 14.1 seeks to prevent and significantly reduce marine pollution of all kinds, particularly from land-based activities and including marine debris and nutrient pollution, by 2025. While these sustainable goals are commendable, they provide little actionable guidance for actors within or downstream from the plastics industry wishing to support these objectives. But ambitious normative visions can play a central role in guiding transitions (Elzen, 2004, p. 56).

Providing a more detailed vision, the Nova Institute suggests it is possible to eliminate the need for virgin fossil-based chemicals and plastics by the middle of the century, replacing them entirely with renewable carbon-based materials.[7] In this scenario, over half (55 per cent) the global demand for plastics would be met through recycling (Figure 3.2) (Nova Institute, 2020b, p. 19). The Nova Institute is not alone in predicting a significant move to recycling. A 2019 report by the consulting firm McKenzie has ambitiously predicted that 50 per cent of the demand for plastics could be satisfied through recycling by 2030 (Hundertmark et al., 2019). At the start of 2021, the CSIRO set out a vision for Australia, also aiming for half the demand for plastics to be satisfied through increased recycling by 2030 (Commonwealth Scientific and Industrial Research Organisation (CSIRO), 2021a, p. 35).

Achieving these ambitious goals is dependent on several assumptions as detailed in the Nova Institute's report. Notably, achieving the recycling target relies on very aggressive adoption of chemical recycling (sometimes referred to as advanced recycling), which includes technologies in the early stages of development (see

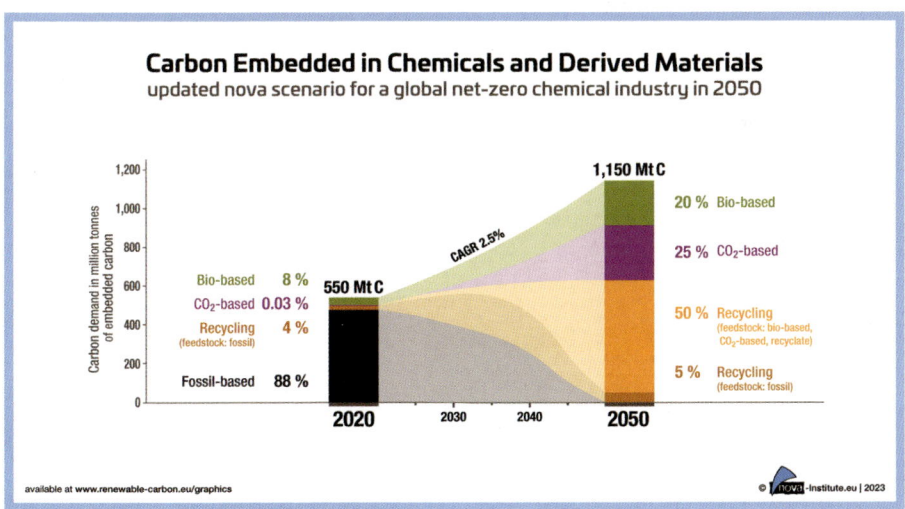

Figure 3.2 Global carbon demand for chemicals and derived materials to 2050. Courtesy Nova Institute.

Chapter 7 for more details).[8] Nova Institute acknowledge that even if it can be demonstrated that chemical recycling is environmentally and commercially viable, and succeeds in attracts sufficient investment to deploy the technology at massive scale, there will still remain a requirement for new sources of carbon to maintain the structural integrity of recyclates and to compensate for material lost during reprocessing. To meet that need, researchers optimistically assume that the required carbon can be extracted from carbon dioxide (CO_2), and through carbon capture and utilization (CCU), eliminate the need for virgin plastic. Again, this assumption is dependent on using technologies that remain undeveloped, or largely unproven at a commercial scale.[9]

It is important to acknowledge the challenges that lie ahead in meeting these recycling targets, as well as the optimistic assumptions that underlie them. However, by 2050, energy and transport regimes will have significantly reduced their CO_2 emissions through investment in renewable power and electric vehicles, decreasing the quantity of carbon embedded in their products. Meanwhile, the petrochemical industry will continue to produce ever-increasing volumes of products, including plastics, which are nearly all based on carbon (Figure 3.3). As the total embedded carbon increases, the regime will face increased scrutiny from those concerned with the preservation of our environment.

An increase in the use of renewable energy is implicit in the projections from the Nova Institute. However, a study by Jiajia Zheng and Sangwon Suh from the

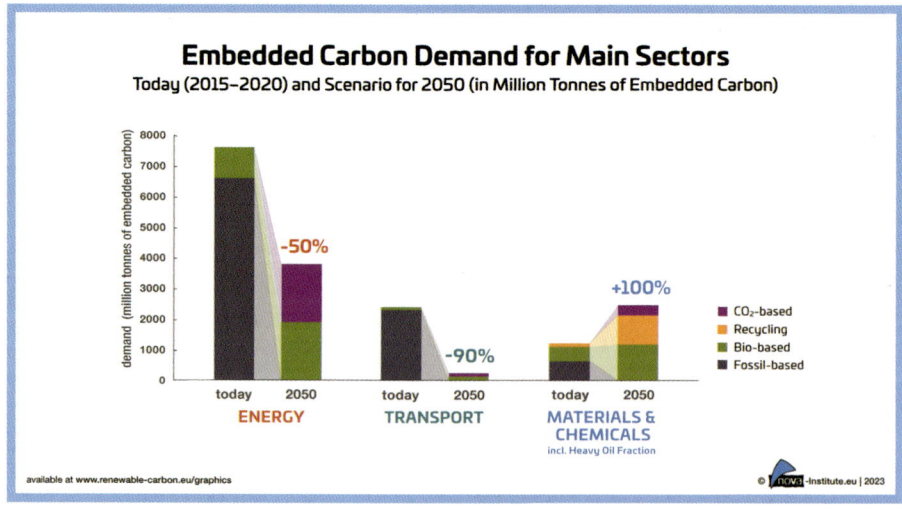

Figure 3.3 Embedded carbon demand by regime. Courtesy Nova Institute.

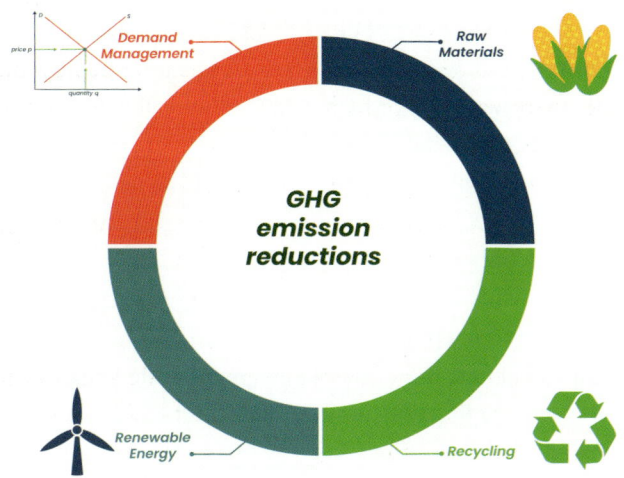

Figure 3.4 Strategies to reduce the environmental impacts caused by plastics. Adapted from Zheng and Suh (2019).

University of California explicitly addressed this issue. The study identified four factors central to reducing the environmental impact of fossil plastics: demand management, energy source, recycling rates and raw materials (Figure 3.4) (Zheng & Suh, 2019, p. 374).

Firstly, the authors suggest that demand for plastics could be managed. This observation is significant as the models previously discussed focused exclusively on supply. By highlighting the role demand plays in constructing this scenario the potential agency of other actors in mapping the future trajectory of the plastics industry, including product manufacturers, designers and consumers is acknowledged. Demand is most easily managed through price. Oil price increases resulting from wars have the potential to stem demand, but wars and price increases are usually, hopefully, short-lived.[10] Carbon taxes or virgin plastic levies are a more permanent solution. Some countries have taken a lead as discussed in Chapter 6, effectively forcing price increases on fossil fuel products while putting downward pressure on the demand for virgin plastics.

Secondly, switching to renewable energy to manufacture plastic would dramatically reduce the GHG impact from the industry, with life-cycle GHG emissions from plastics reduced by between 51 per cent and 62 per cent (Zheng & Suh, 2019, p. 375). While the Nova Institute's model assumes an overall shift to renewable power, no specific reference is made to the impact of this change on GHG emissions from the production of plastics. The Zheng and Suh model highlights the impact the source of energy used to make plastic has on its overall environmental impact of plastic. Decarbonizing the energy system is a long-term solution requiring political will and serious investment and, ironically, many plastic components. The fact that petrochemical facilities produce heat as a biproduct which can be used to generate power at low cost also disincentivizes change. Some progress is being made, with DuPont taking the lead by targeting 60 per cent renewable energy by 2050 and reporting that 20 per cent of its electricity supply came from renewable sources in 2020.[11]

Designers do not have agency over these first two strategies to reduce demand for fossil plastics; however, they are closely involved with the remaining two: increasing the use of recycled plastics and bioplastics. Increasing recycling rates reduces the demand for virgin fossil fuels. If the proportion of plastics recycled increased steadily to 100 per cent by 2050 (which the authors acknowledge to be unrealistic), life-cycle GHG emissions would decrease by 25 per cent but that could reach 77 per cent if renewable energy is used for reprocessing. The authors emphasize the synergistic benefits of using renewable energy to power increased recycling activity (Zheng & Suh, 2019, p. 375).

Finally, the use of bioplastics can be increased. In common usage, 'bioplastics' is often assumed to signal the material has been made from 100 per cent renewable organic sources. However, the term is often used to describe a wide range of materials; some are entirely made from renewable organic feedstock and others (sometimes called bio-based) often only contain a percentage of renewable organic contents mixed with fossil fuels. Bio-based plastics are not necessarily compostable or biodegradable. Biodegradable and compostable plastics biodegrade in certain specific conditions. Confusingly, some biodegradable plastics are made from fossil fuels. In the Zheng and Suh study, only bioplastics made from corn or sugar are included and the impact on agricultural land use is considered. This led them to conclude that, in terms of GHG life-cycle emissions, recycling conventional plastics may be equally beneficial as switching to bioplastics (Zheng & Suh, 2019, p. 375). This analysis fails to investigate the GHG impact of higher generation feedstocks (such as algae and CO_2), which, when used to make bioplastics, are likely to have different environmental profiles.[12] Despite these limitations, the authors conclude that the best GHG emissions scenario from their analysis can be achieved when plastics are all derived from sugarcane using 100 per cent renewable energy. In this unlikely scenario a 93 per cent reduction in GHG emissions could be achieved.

The bioplastic market is embryonic, accounting for less than 1 per cent of all the plastic produced in 2023, but it is growing fast with a projected CAGR of over 17 per cent through to 2030.[13] This forecast might well be superseded by more aggressive growth. At the end of 2020, Greenpeace International reported that thirty-six Chinese producers had planned or built new biodegradable plastic manufacturing facilities. These new investments will add more than 4.4 million tonnes of production capacity per year, representing a sevenfold increase in Chinese production of bioplastics (Greenpeace International, 2020, p. 11). In November 2022, Braskem announced plans to increase production of biopolymers produced from sugarcane in Brazil from 200,000 tons to 1 million tons by 2030 (Braskem, 2022). Despite those gains and given the expected increase in demand for plastics in the coming decades, the role bioplastics can play in transitioning toward renewable carbon sources for plastics is likely to remain limited in the period to 2050.

After comparing the potential GHG savings that could be achieved by adopting a mix of the four strategies detailed above the authors concluded that the only way to achieve an absolute reduction in GHG emissions is through demand management (reducing industry growth to 2 per cent per annum) in addition to a combination of the other three strategies (Zheng & Suh, 2019, p. 377). Zheng and Suh highlight the enormous magnitude of the changes the petrochemical industry must undertake to limit growth in GHG emissions in the face of burgeoning demand.

Another scenario with a focus on the demand side has been developed by the Ellen MacArthur Foundation. By the end of 2020, the Foundation had secured commitments from more than five hundred major FMCG organizations, accounting for 20 per cent of all the plastic packaging used, to eliminate the use of plastics where possible or move to reusable, recyclable or compostable packaging by 2025 (Ellen MacArthur Foundation, 2020, p. 7).[14] Addressing the aims of UN goal 12.5, the Ellen MacArthur Foundation has been successful in publicizing the concept of a circular economy where 'business models, products, and materials are designed to increase use and reuse, replicating the balance of the natural world where nothing becomes waste, and everything has value' (Ellen MacArthur Foundation, 2022). The Foundation's work is focused on the important challenge of reducing the use of virgin plastics in packaging, which accounts for the largest proportion of virgin plastics consumption. However, the remaining 64 per cent of plastics used for purposes other than packaging are not the target of these efforts.

None of the visions included in the Ellen MacArthur Foundation report offer a solution to the problems caused by our ever-increasing consumption of plastics, but a framework to guide decision-making by those concerned with reducing the environmental impacts caused by plastics is provided. These narratives serve to challenge the entrenched regime and emphasize opportunities arising from alternatives. The visions expressed by the Foundation and other similar organizations are driving interest in recycled plastics and bioplastics by highlighting the need to reduce demand for virgin polymers, as an important step toward developing a more sustainable relationship with plastics.

Significantly, visions expressed by organizations dedicated to tackling environmental issues highlight the potential agency of actors outside the petrochemical industry in driving change within the plastics sector. While the scale of change needed for a more sustainable relationship with plastics may seem overwhelming for individual actors, it may not require a complete reduction in demand or the complete elimination of growth projections to disrupt the industry. The Carbon Tracker Initiative, a London-based not-for-profit think tank, suggest that just by making changes at the margin, reducing the CAGR for virgin plastics from 4 per cent to 1 per cent, will be sufficient to cause significant disruption to the market, and strip the fossil fuel industry of much of its forecast profit growth (Carbon Tracker Initiative, 2020, p. 18). The petrochemical industry is heavily reliant on continued growth in demand for virgin plastics to meet profit projections, as demand for fossil fuels slows and eventually declines. Any slowdown in the plastics market has the potential to negatively impact growth targets and shareholder returns. Investments in infrastructure to meet the anticipated growth in demand and designed to be productive and deliver returns over decades risk being written off as stranded assets. This analysis highlights the important role consumers, designers and product manufacturers can play in impacting demand without relying on government intervention. Reducing packaging, improving product life expectancy or giving preference to refillable containers are given as example of the agency held by these three actors respectively (Carbon Tracker Initiative, 2020, p. 23). Avoiding the use of virgin plastics altogether, by replacing them with renewable carbon-based materials, will potentially exert an even bigger impact on the industry's growth plans.

a. The importance of design

Given that consumer products account for just 11 per cent of plastic usage, it might be more productive to focus on reducing the use of virgin plastic elsewhere, particularly by in packaging, which accounts for 36 per cent (Figure 3.5). Several

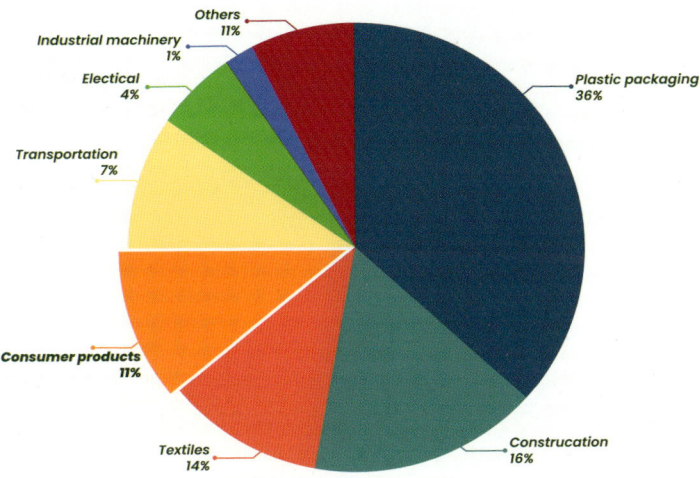

Figure 3.5 Breakdown showing where plastic was used during 2015 – 11 per cent of plastics were used to make consumer products. Adapted from Geyer et al. (2017a).

initiatives, including those led by the Ellen MacArthur Foundation, are already underway to focus on that problem. As the amount of virgin plastic used for packaging decreases, the persistent use of traditional plastics in consumer goods is likely to attract more criticism, as the evidence of its negative impact on the environment and our health continues to grow. It is almost certain that product designers, together with clients and manufacturers, will face increased scrutiny as they participate in the production of an endless array of consumer goods, many of which are designed to be quickly disposed of simply to make way for the next 'improved' model.

Incorporating renewable plastics into consumer products will positively influence the speed with which these markets develop. Designers can capitalize on the shifts in consumer sentiment away from fossil plastics and design consumer products using more environmentally friendly renewable carbon-based materials. Incorporating these materials in consumer products develops an important niche market that will support the expansion of renewable polymer production. Branching the use of renewable plastics into new market segments increases demand and helps develop much needed economies of scale for material manufacturers (F. W. Geels, 2005b, pp. 59–60). Expanding the renewable plastic market to include consumer products is also likely to disproportionately increase its value, as higher quality materials (delivering more profit for material manufacturers) will be needed for consumer products designed for longevity. Potential for higher profit margins is particularly significant to encourage advances in bioplastics, as current markets are all relatively low-value (packaging, agricultural film and mulch, disposable plastics bags and disposable plastic tableware) (Grand View Research, 2022a). Increasing both the size and value of the market will attract more investment to bioplastics, hastening the development of the industry.

3. Summary

There is a growing body of evidence to demonstrate the long-term negative environmental and health impacts caused by the production and disposal of plastic. Current industry forecasts for continued expansion of fossil plastic production are unsustainable. Alternative realistic scenarios of how we can meet the growing demand for plastic more sustainably are being developed but these all depend on a range of strategies to drive the rapid development and adoption of renewable plastics. Importantly, both supply and demand must be generated to drive the market for renewable plastics. Supply can be increased through the expansion of mechanical recycling activities and development of the chemical (or advanced) recycling industry. Designers and manufacturers have a crucial role to play in increasing demand and accelerating the transition toward renewable plastics, by saying no to virgin plastics and choosing to work with bioplastics and recycled plastics. I return to the topic in the final two chapters of this book to provide practical advice on how designers and manufacturers can encourage and support the transition towards renewable plastics. However, with an ever-increasing range of recycled plastics and bioplastics becoming available the first challenge is to identify which materials are most suitable use when designing consumer goods and deliver the best environmental outcomes.

4 Identifying environmentally responsible products

Since early in the twenty-first century there has been growing interest from both designers and manufacturers in experimenting with recycled plastics and bioplastics (renewable plastics). The primary motivation to work with these materials is often to reduce the environmental impact of a product. But how can clients and designers be sure the products they are developing achieve this goal? What can be learnt and what is worth replicating from early experiments with renewable plastics? Complicating any comparison of environmental performance is the fact that components made from a variety of materials are often sourced from (frequently opaque) global supply chains and assembled into products that are distributed globally.

To examine the environmental credentials of various renewable plastics I searched for a product category where there has been plenty of activity using these materials. I limited my search to product categories that are unlikely to consume energy or significant quantities of other resources during use, allowing consideration of the use phase of a product's life cycle to be avoided. I decided to focus on chairs for reasons detailed in the Introduction. Significantly, chairs are often among the first products commissioned to showcase a new material or technology for use in product design.

I limited my research to chairs where plastic is used to provide support for the sitter i.e. the legs and/or the seat and back of the chair must be made from plastic. Hence, a chair comprising a plastic shell resting on metal legs could be included in my study while a design completely covered in plastic, while relying on another material to provide support would not qualify. Similarly, office chairs dependent on metal frames to provide support (for example, Aeron chairs) do not qualify for inclusion. This somewhat arbitrary decision has meant that some promising designs that feature alternative materials, such as Vepas' *Hemp-Fine Chair*, with a shell made from hemp and bio-resin resting on a metal cradle, were excluded from the analysis. Only designs that are (or were) commercially available are included; proof-of-concept pieces and other prototypes are also excluded.

I needed a methodology to compare the environmental performance of chairs made from renewable plastics using only the limited information published by manufacturers. I developed a simplified eco-audit tool to evaluate factors such as resource use and transportation and waste impacts, to determine the overall environmental performance of each chair. This tool is then used to identify a handful of designs that deliver the most superior environmental performance. While my focus here is on plastic chairs many of the same principles apply when comparing the environmental performance of plastic products within a given product category that does not consume resources during use. The analysis highlights the factors that should be considered when selecting a renewable plastic for a design project. Many of the findings from this analysis are of relevance to all those working with plastics to create consumer goods.

1. Renewable carbon plastic chairs

Jane Atfield experimented with making chairs from recycled plastics (HDPE) as far back as 1992, when a small batch of RCP2 Chairs were produced (Figure 4.1), with additional small batches released since (Fiell & Fiell, 2009, p. 228). However, it is only in the twenty-first century that designs using renewable plastics have begun to regularly feature in mainstream large-scale production. Designs are being made from recycled plastics (post-industrial, post-consumer and ocean plastics), bioplastics (derived, at least partially, from renewable organic material), or natural fibres, each with their own unique set of environmental impacts.

I identified sixty chair designs manufactured using renewable plastics between 2007 and 2022 (many of these chairs are illustrated in the following chapters). Each of these designs is effectively a niche experiment with renewable plastics. Some of those chairs have been manufactured using relatively new technologies that promise to reduce the material and/or energy required to manufacture products, potentially further impacting environmental performance. Just over half the chairs (thirty-three) are monoblocs. Many of the remaining designs have been developed by combining plastics with other more traditional materials such as

Figure 4.1 *RCP2 Chair*, Jane Atfield, 1992. Courtesy Jane Atfield.

wood or metal, further complicating attempts to measure and compare the environmental profile of each design. Without an accurate assessment it is impossible to know if these niche experiments have succeeded in delivering more sustainable seating solutions.

a. Life-cycle assessment

Life-cycle assessments (LCAs) are the most sophisticated methodology for evaluating the environmental impacts of products over their entire life cycle. However, LCAs are onerous, time-consuming and expensive to produce, ideally requiring practitioners with expertise in thermodynamics and chemistry. As a results LCAs are often simply ignored by many industries. In January 2023, I found only three LCAs on the websites of the thirty-four furniture manufacturers responsible for the sixty chairs examined in this study. With so few LCAs available I developed an alternative simplified approach to enable a comparison of environmental impacts. However, before moving on to describe my approach, it is worth noting that while LCA results are often presented as a definitive evaluation of environmental performance there are significant limitations and challenges to the methodology which are worthy of consideration, especially when attempting to compare results from different sources.

LCAs were developed to provide a systematic framework to quantify and assess various environmental indicators such as greenhouse gas emissions, energy consumption and resource depletion. A comprehensive evaluation of the environmental efficiency of a product can be calculated across multiple impact categories including global warming potential (GWP), resource depletion, acidification potential, eutrophication potential, photochemical oxidant formation, ozone depletion, ecotoxicity, human toxicity, particulate matter formation, land-use, energy and water consumption, among others. Various frameworks and guidelines are used to calculate these results which are complex and difficult to communicate. To simplify messaging various methodologies have been developed to weight results, highlighting some impact categories while downplaying or ignoring others and thereby introducing more variables.

LCAs can only be calculated when highly detailed and accurate information about a product becomes available, often only after a product has entered production (impeding the relevance of the tool as an aid to designers). Precise quantities and specifications of all materials used are required, together with details of resources consumed during their extraction and processing. To simplify this process, precalculated databases are commercially available and can be accessed to attribute data for most inputs. However, proprietary data restrictions often hinder transparency and consistency across data providers, preventing practitioners from verifying the accuracy and relevance of selected data. LCAs also require detailed information of the energy and additional resources consumed during manufacture of the final product, including consideration of any waste impacts. Often this information is obtained from published literature which may be of variable quality or calculated using undefined assumptions. As a result, great care should be taken when directly comparing LCA results, as this can only be done when precisely the same methodology has been observed, and full details of assumptions and sources of data are disclosed.

Attributional LCA involves the delineation of system boundaries to define the scope of analysis. Manufacturers often restrict the boundary of their analysis as 'cradle-to-gate' or 'gate-to-gate', as their primary focus is to improve their procurement and manufacturing processes. However, this approach ignores impacts caused by products after they have left the factory gates, which manufacturers argue they have little control over. A full 'cradle-to-grave' analysis attempts to measure a product's environmental impacts from extracting raw resources from the earth, through production, accommodating recycled content, distribution, use and disposal/reuse or recycling at end-of-life. It is reasonable to argue that a manufacturer cannot determine, for example, precisely what will happen to a product at the end of its useful life. However, limiting LCAs to 'cradle-to-gate' obfuscates the responsibility of the manufacturer to minimize any potential end-of-life environmental impacts (for instance, by selecting recyclable materials and ensuring products can be easily dismantled). Cradle-to-gate LCAs also fail to reward well-designed, well-made goods destined to last a long time and thereby spreading all impacts over a longer functional life and delaying any end-of-life impacts.

Additional complexities arise when calculating LCAs for products with plastic components. To calculate the impacts caused by the production of plastics, LCA practitioners reference secondary data sources to simplify calculations, but this ignores differences in the plastic production processes and prevents accurate quantification of GHG emissions from the plastic supply chain (Pires da Mata Costa et al., 2021).

An accurate LCA for a plastic product or component requires an understanding of the feedstock and processes used to create the material, as this is often the biggest contribution to the environmental impact of a plastic product that does not consume resources during use. While most plastics are made from oil, North American and European plastics are increasingly likely to have been made using gas. Plastic (particularly PP) from China (where production was increasing by nearly 6 per cent per annum prior to Covid-19) is increasingly more likely to have been made from coal, due to an aggressive programme of investment designed to make China more self-sufficient and support coal miners impacted by the transition to renewable energy (Chang, 2016). Each feedstock has its own environmental profile, with coal to olefin technologies emitting up to triple the CO_2 emissions produced by oil to plastic production and consuming 4.2 times as much water (Greenpeace, 2017; HSBC Global Research, 2014, p. 16). Commercially available datasets quantifying the environmental impact of commonly used plastics overlook these complexities and frequently fail to be transparent about the assumptions used, thereby failing to encourage the selection of less environmentally damaging variants.

Even within a single category of feedstock, there are significant variations in impact estimates depending on the data source referenced. A comparison of fourteen datasets modelling the GHG emissions caused by refining petroleum in Europe found up to 300 per cent variation in the results for major refined products including naphtha, the oil-based feedstock used to make plastic (Johnson & Vadenbo, 2020). Note this study focused on a single impact category but the same methodological discrepancies affect all impact categories to varying degrees.

In early 2024, one prominent LCA data provider made changes to its methodology which resulted in a 107 per cent increase in the estimated carbon footprint of naphtha (Renewable Carbon Initiative, 2024). When these data sources are used to subsequently calculate the environmental impact of products that include plastics made from oil these variations are carried through. GHG emission variations are likely to be further magnified, as emissions typically get weighted highly when summarizing LCA results.

When considering bioplastics, the analysis of feedstocks become more complex, as the variety of feedstocks used is broad, with many potentially impacting the ecosystem at end-of-life due to the use of genetically modified organisms (GMOs), pesticides and/or fertilizers. All these impacts, together with the consumption of land and water, need to be considered in a detailed LCA analysis of products made from bioplastics (Piemonte & Gironi, 2011). As material manufacturers operating in this fast-developing industry are particularly keen to protect their intellectual property, accessing the required level of detail for many of these materials is particularly challenging, if not impossible.

Nigel Howard, owner of Clarity Environment, has gained decades of experience working with LCAs in the UK, United States and Australia. Since migrating to Australia, Howard managed the Building Products Life Cycle Inventory project to establish a national consensus methodology and database of consistent LCA data for the building materials and products industries. While working for the Infrastructure Sustainability Council of Australia, he developed the LCA-based Materials Calculator and the Materials and Resources credits for the Infrastructure Sustainability IS rating tool. Howard, who kindly agreed to review this section of this book, is a vocal critic of the methodological inadequacies of LCA applications and actively campaigns to address these issues.

As LCA results are presented as hard data evaluating the environmental impacts of products across multiple impact categories it is tempting to regard these outputs as irrefutable facts. International standards (ISO 14040/4) have been developed specifically to guide consistency across LCAs. Howard argues that these have failed to 'evolve sufficiently since their original authorship in 1997 (and revision in 2006) to be considered fit-for-purpose for standardizing LCA outcomes. As a result, LCA methodology is interpreted differently' or gamed by practitioners facilitating the selective production of 'winning stories' for commissioning clients (Howard, 2017, p. 3). Attempts to standardize methodologies so that consistent and replicable results can be produced are frequently stymied by commercial vested interest groups. Standardization would ensure consistency of results (between and within the supply chains of all product systems and processes), which are comparable, repeatable and traceable back to original authors documentation and provide a truly legitimate basis for decision-making. Howard concludes that until there truly is a single consistent methodology that is universally adopted LCA will continue to fail to have any meaning. Due to this lack of standardization, LCAs, provided by different manufacturers cannot be directly compared.

I assumed it would still be a relatively simple task to locate sufficient data to make an accurate comparison of the environmental impact of similar products. I expected product specifications detailing quantities of materials and manufacturing processes to be readily available. I also anticipated that major manufacturers would publish sustainability reports or environmental impact statements, outlining environmental achievements and aspirations. With a few exceptions, surprisingly little information is in the public domain. Only four out of the thirty-four manufacturers had produced a standalone sustainability report at the beginning of 2023, with only three publishing full LCAs as noted above. Surprisingly, only seven product manufacturers made any reference to using, or planning to use, renewable energy in their production facilities. Despite producing chairs using renewable plastics, four of the companies surveyed failed to identify any sustainability initiatives beyond using recycled materials.

Manufacturers typically publish generic statements about their sustainability commitments or achievements, without specifying actions or committing to targets at the organizational level, let alone provide detailed information at the product level. With access to such limited information on specific products, how can environmentally conscious consumers make informed purchase decisions? How can design practitioners easily identify which combinations of materials and manufacturing processes are most likely to assist in reducing the environmental impact of their creations? To answer these questions, I developed a simple tool to enable a comparison of the environmental impacts of designs with the constraints of the limited information available from manufacturers websites.

2. Developing an eco-audit tool

'The waste hierarchy' was developed to evaluate waste management by ranking the environmental impact of processes from most favourable to least favourable. I adapted this tried and tested tool, popular with design academics in the 1980s (the early years of green design), to assess its relevance to designers and manu-facturers who are keen to evaluate the environmental profile of their creations (Mackenzie, 1997; Madge, 1997, pp. 45–54).

The accuracy of the rankings within the hierarchy applied specifically to plas-tic waste (see Figure 4.2), were verified by research conducted by the Waste and Resources Action Programme (WRAP) in the UK. A detailed comparison of eight LCA studies identified mechanical recycling as the preferred disposal alternative for plastics when compared for climate change impact, depletion of natural resources and energy consumption (WRAP UK, 2010). Recycling delivers both energy security and environmental benefits. The amount of fossil feedstock required for production is lessened and, more importantly, recycling requires less energy and emits fewer GHG emissions. On average recycling one tonne of plastic saves about 0.6 tonnes of CO_2 compared with landfilling (Bennett, 2012a). That is why, from an environmental perspective, recycling is the preferred end-of-life option for plastics.[1]

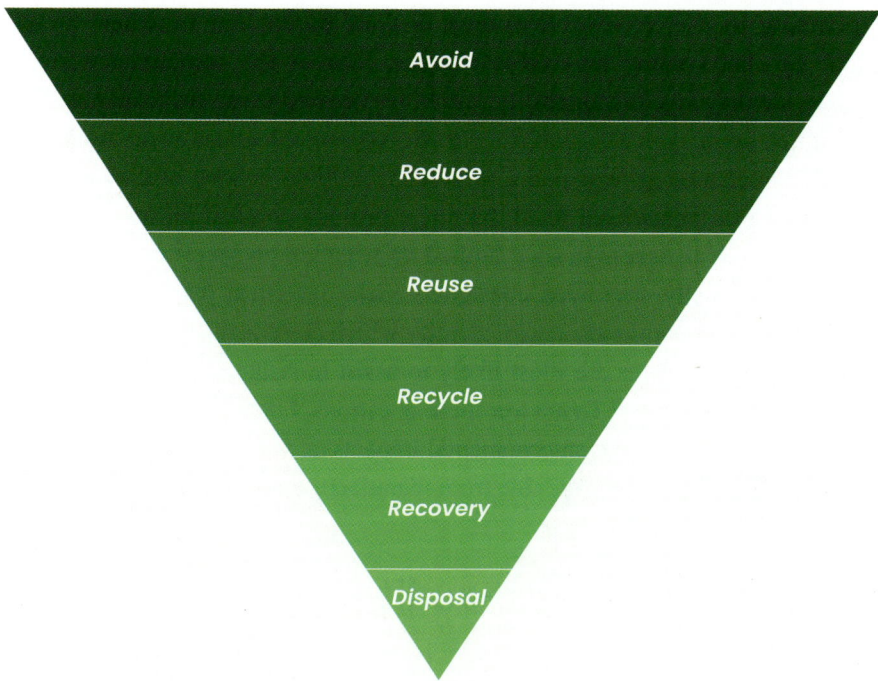

Figure 4.2 The waste hierarchy – avoiding waste altogether is the preferred option with disposal the least preferred option.

In the 1990s, it became fashionable for manufacturers to label their products, including chairs, as being recyclable. But a label is not a guarantee of what will happen to the product at the end of its life. I have therefore differentiated between designs that are theoretically recyclable from those actually made from recycled plastic, with reuse the more preferable outcome.

Since 2007, there have been a growing number of innovations aimed higher up the waste pyramid (Figure 4.3). These new approaches focus on designs made from recycled plastic, manufactured using new technologies developed to reduce the amount of energy or material required, or incorporate alternative materials to avoid the use of petrochemical plastics altogether. I have modified the waste hierarchy to reflect those innovations. The most significant difference is in the definition of 'avoid'. In the original waste hierarchy 'avoid' is defined as avoiding waste all together by encouraging the reduction of the amount of virgin materials extracted and used. In my adaptation, to guide designers working with plastics, 'avoid' is redefined as 'avoiding the use of fossil-based plastics'. While 'reuse' includes all activities involving using recycled plastics and 'reduce' includes all efforts to minimize the use of resources included in the design or consumed during production. Appling this analysis to the sixty chairs using plastics made from renewable carbon delivered the results shown in Table 4.1.

The waste hierarchy is a useful tool to guide material selection toward more sustainable outcomes. However, it is not sufficiently detailed to evaluate and

Figure 4.3 Preferred solutions from the waste hierarchy.

Table 4.1 Waste hierarchy analysis of renewable plastic chairs

CHAIR	DESIGNER	MANUFACTURER
Avoid – Natural fibres		
Flax chair	Christien Meindertsma	Label/Breed
Hemp chair	Werner Aisslinger	Moroso
Jin	Jin Kuramoto	Offecct
Coastal Furniture	Nikolaj Carlsen	TangForm
Avoid – Bioplastics		
Jet and Paco	Tramontina	Tramontina
Kuskoa Bi	Ander Lizaso & Jean Louis Iratzoki	Alki
Siamese Chair	Karim Rashid	A Lot of Brasil
Albert Kuip Coffee Chair	Studio APE	Zuiver
Avoid and reduce – Bioplastics (3D printed)		
Voxel	Manuel Garcia	Nagami
Puzzle Chair	Joris Laarman	Joris Laarman Lab
Reduce – 3D printing		
Chubby Chair	Dirk Vander Kooij	Dirk Vander Kooij
Reduce – Artificial intelligence		
AI Chair	Philippe Starck	Kartell
Recycled – Ocean plastics		
DuraOcean	ScanCom	ScanCom
Ibiza (chair)	Eugeni Quitllet	Vondom
Love	Eugeni Quitllet	Vondom
Ocean OC2 Chair	Jørgen & Nanna Ditzel	Mater
S-1500	Snøhetta	Nordic Comfort
Ocean Chair	APE Studio	Zuiver
Wave	Polywood	Polywood
Louvre	Danny Fang	Kain
Brooklyn Chair	Eugeni Quitllet	Vondom
Africa Chair	Eugeni Quitllet	Vondom

DATE	MATERIAL 1 (R= RECYCLED CONTENT)	MATERIAL 2
2015	flax	PLA
2011	hemp (70%)	water based resin
2018	flax fibre	bio-resin
2019	seaweed	wood (legs)
2018	renewable PE	
2015	PLA (hybridized)	wood (legs)
2014	plant based plastic	aluminium legs
2021	coffee waste, PP (57.5%)	wood (legs)
2016	PLA	
2014	PLA	
2012	recycled refrigerator plastic (R96%)	
2019	thermoplastic technopolymer (R100%)	
2020	recycled fishing industry waste	plastic debris
2019	ocean plastic (PP)	fibreglass
2018	ocean plastic (PP)	fibreglass
2018	ocean plastic (PP – 960 g)	steel (legs)
2019	ocean plastic (PP – 1,500 g)	steel 20% recycled (legs)
2022	ocean plastic (PP, R100%)	steel (legs)
2020	ocean plastics (HDPE)	
2013	ocean plastic (PP)	
2019	ocean plastic (PP)	fibreglass
2019	ocean plastic (PP)	fibreglass

(continued)

Table 4.1 Waste hierarchy analysis of renewable plastic chairs, *continued*

CHAIR	DESIGNER	MANUFACTURER
Pedrera	Eugeni Quitllet	Vondom
Recycled – Small-scale projects		
Butter Seat	Nicholas Karlovasitis & Sarah Gibson	DesignByThem
Charlie Chair	Vanessa Yuan & Yoris Vanbriel	ecoBirdy
Meltdown Chair	Tom Price	Tom Price
Recycled mixed		
Be Pop	Ludovica + Roberto Palomba	Kartell
Re-Chair	Anronio Citterio	Kartell
Masters RE	Philippe Starck & Eugeni Quitllet	Kartell
Lady B Go Green	Studio Zetass	S.CAB
Ginevra Go Green	SCAB Design	S.CAB
Hug Go Green	Meneghello Paolelli	S.CAB
Sai Go Green	SCAB Design	S.CAB
Pip-e Green	Philippe Starck & Eugeni Quitllet	Driade
Nemo Green	Fabio Novembre	Driade
Roly Poly Green	Faye Toogood	Driade
Sissi Green	Ludovica e Roberto Palomba	Driade
Recycled PET		
111 Navy Chair	Emeco	Emeco
Nobody	Boris Berlin & Poul Christiansen	Hay
On & On Stacking Chair	Barber Osgerby	Emeco
Recycled post-consumer polypropylene		
Summa Range (Sissi Chair & Diana)	Tramontina	Braskem and Tramontina
Falk	Thomas Pedersen	Houe
N02 Recycle	Oki Sato	Fritz Hansen
Fluit	Archirivolto Design	ACTIU

DATE	MATERIAL 1 (R= RECYCLED CONTENT)	MATERIAL 2
2019	ocean plastic (PP)	fibreglass
2011	HDPE (R80%)	
2018	recycled toys	
2007	PP	
2022	thermoplastic technopolymer	
2022	thermoplastic technopolymer	
2022	thermoplastic technopolymer	
2020	technopolymer (R60%)	steel base
2020	technopolymer (R60%)	
2022	technopolymer (R60%)	
2020	technopolymer (R60%)	
2021	mixed industrial & consumer	
2021	mixed industrial & consumer	
2021	mixed industrial & consumer	
2021	mixed industrial & consumer	
2010	PET (R100%)	fibreglass (35%)
2007	PET felt (R100%)	
2019	PET (R100%)	fibreglass (20%) non-toxic pigments (10%)
2019	PP	
2019	PP (75%)	fibreglass 20% + wood (legs)
2019	PP (R100%)	steel (legs – 50%)
2022	PP (agricultural sector)	fibreglass recycled (20%)

(continued)

Table 4.1 Waste hierarchy analysis of renewable plastic chairs, *continued*

CHAIR	DESIGNER	MANUFACTURER
About a Chair Eco	Hee Welling	Hay
Recycled post-industrial polypropylene		
Broom Stacking Chair	Philippe Starck	Emeco
Green Chair	Javier Mariscal	Mobles114
Odger	John Löfgren & Jonas Pettersson	IKEA
1 Inch Reclaimed	Jasper Morrison	Emeco
Bell Chair	Konstantin Grcic	Magis
Fibre Chair	Iskos-Berlin	Muuto
Max	Christop Jenni	Maxdesign
Tip Ton RE	Barber Osgerby	Vitra
Elemental	Ronan & Erwan Bouroullec	Hay
Juno 02	Studio Irvine	Arper
Rely Chair	Hee Welling	&tradition
Adell	Lievore + Altherr Désile Park	Arper
Recycled mixed polypropylene		
Babila XL	Odo Fioravanti	Pedrali
Remind recycled	Eugeni Quitllet	Pedrali
Move	Formway	Noho

Key:

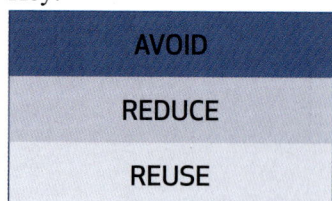

DATE	MATERIAL 1 (R= RECYCLED CONTENT)	MATERIAL 2
2020	PP	wood (legs)
2012	PP (R75%, since increased to 90%)	wood (15%)
2016	PP (R100%)	wood or metal bases
2017	PP (70%)	wood (30%)
2018	PP (88%)	wood (2%)
2020	PP (R=almost all)	fibreglass
2014	PP (R80%)	wood (25%) steel, wood or aluminium legs
2018	PP	fibreglass
2020	PP	
2018	PP (80%)	fibreglass (20%)
2022	PP (70%)	ABS support
2022	PP (90%)	fibreglass (10%) steel
2020	PP (80%)	wood or steel bae
2021	PP post-industrial (50%)	polypropylene consumer (50%), steel base
2021	PP post- industrial (50%)	polypropylene consumer (50%)
2020	PP	Nylon 6 (from fishing nets)

compare designs within each category. Additionally, there is no guarantee that material selection alone defines the overall environmental profile of a product.

3. Environmentally Responsible Product Rating (ERPR)

A more detailed method was required to identify the most environmentally friendly designs. To address this need, I developed a simplified eco-audit tool that assigns an Environmentally Responsible Product Rating (ERPR) to each design. ERPR is a simplified version of a tool originally developed by Yale Professor, Tomas Graedel (Graedel, 1998). My version of the tool differs from this and other more complex eco-audit offerings by focusing on the five most important considerations for actors involved with the creation of products that do not consume energy resources during use (Figure 4.4).

Graedel's simplified LCA included consideration of the use phase, but this is not relevant for most plastic chairs (where maintenance is often limited to the occasional wipe with a damp cloth) and has therefore been excluded from the analysis. It should be noted that the serviceable life of some of the chairs made from multiple materials might be extended through maintenance or replacement of parts, but I have ignored these possibilities to simplify the analysis.

Importantly, my ERPR tool differentiates itself from other eco-audit methodologies, including Graedel's original ERPR tool, by attempting to evaluate the appeal of a product. Appeal is defined here as the attributes that command atten-

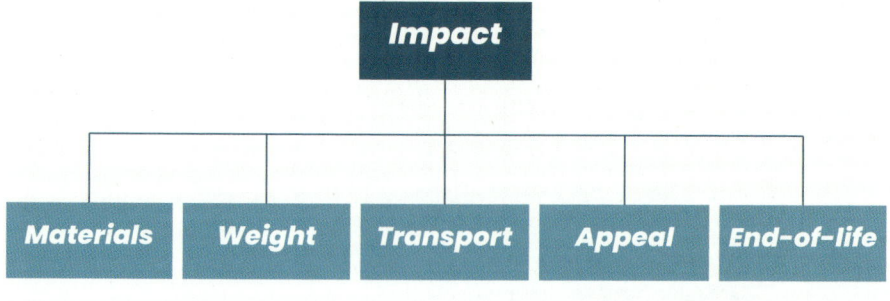

Figure 4.4 Attributes of plastic chairs evaluated by the eco-audit tool.

tion and persuade people to purchase a product. It encompasses the aesthetic characteristics of a chair, such as its style, texture, appearance and comfort level. These characteristics are largely subjective but play a vital role in determining sales. If an environmentally friendly product fails to appeal to the market, it will, most likely, be substituted by a less-sustainable alternative.

If renewable plastics are deemed unattractive, purchasers (including consumers and business purchase decision makers) will continue to select virgin fossil versions of the same products. The designer can play a critical role in developing an appreciation for the unique aesthetic qualities of renewable plastics. The environmentally conscious designer must extend their concern from a focus on production to recognize that the act of consumption is inherently intertwined with the act of creation. This is particularly important for products with relatively inelastic demand, such as chairs. Once a consumer has identified a need for a chair, they are likely going to fulfil that need by making a purchase; conversely, they are unlikely to impulse buy more chairs than needed just because they like their appearance.

Many factors that contribute to the appeal of a design can only be assessed subjectively. I have restricted my analysis to three easily quantifiable criteria crucial to the appeal of a chair: its price, the range of colours available and the number of variants available (Figure 4.5). Lower-priced chairs are likely to appeal to a wider audience, while those offered in a wide range of colours are more likely to find favour with end-users searching for seating solutions that match their interior décor. Finally, if a chair can be adapted to suit the varying needs of different niche markets, potential sales can be increased significantly, just by gaining a small market share in each market segment. By offering a variety

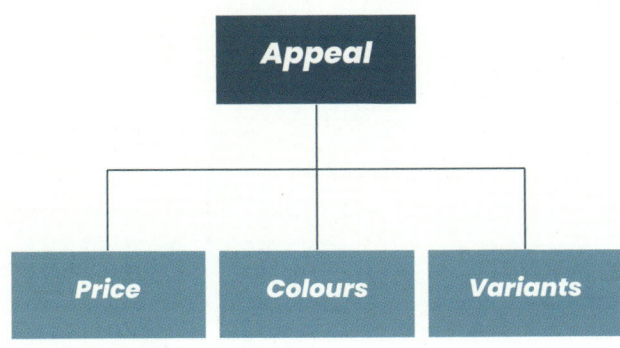

Figure 4.5 Easily quantifiable factors that contribute to the appeal of a design.

of bases and presenting shells with or without arms, a design can be adapted to suit a variety of uses across the residential, business and contract segments (Figure 4.6).

It can be argued that other less-quantifiable attributes of appeal, such as comfort, feel and quality of detailing, are more important in the selection of a chair. More sophisticated techniques to evaluate these dimensions of appeal are needed to measure attributes such as these. Despite this, by including easily quantifiable elements of appeal, I hope to emphasize the need to create sustainable designs that are attractive to their intended audience. Products that do not sell, regardless of their eco-credentials, simply waste resources. By focusing on both environmental impacts and appeal, designers can create products that are not only more sustainable but also commercially viable.

The choice of materials is particularly significant for products made entirely from plastic, as the energy consumed in creating plastic will almost certainly dwarf the environmental impacts caused by the product throughout its lifespan. Consequently, the amount of material used is a crucial factor in a product's environmental performance. Without access to detailed information about the

Figure 4.6 *Falk*, Thomas Pedersen for Houe, 2019. Originally available with five base choices and two shell options to suit a variety of domestic and corporate needs. Coloured pads are available in a range of colours to improve appeal, compensating for the lack of colour choices for the shell. Courtesy Houe.

processes and resources used during production, the final weight of the product is assumed to be a surrogate indicator of the efficiency of these operations. Using less material results in lower energy needs for processing.[2] The efficiency with which products are distributed is also significant – developing chairs that can be stacked or dismantled for transport lowers distribution impacts. Even where a full LCA analysis is conducted these impacts are often overlooked, as manufacturers frequently only measure activities which take place on their premises (cradle-to-gate). Similarly, what happens to products at the end of their useful lives can have a big environmental impact and is ultimately determined by the end-user. While the choice of material primarily determines the end-of-life prospects of any product, designing for disassembly and ensuring that all components are permanently labelled to indicate their recyclability (especially important for plastic components) can promote improved outcomes.

For each of these four factors (materials, weight, transport and end-of-life), a score ranging from zero to four is awarded based on information readily available from manufacturers' websites. More details of the scoring system adopted can be found in the Appendix in this book. In summary, categories were established by inspecting the data, with many of the category definitions determined by dividing the sample into five roughly equal segments or quintiles. The creation of categories is therefore somewhat arbitrary, for example, a chair weighing 2.9 kilograms will receive the top score of four points, while a chair weighing 3 kilograms will receive only three.

Consideration was given to weighting scores; however, this would introduce another level of subjectivity to the results. For example, the end-of-life prospects for chairs might be considered less critical, as most chairs are designed to be long-lasting and withstand years of continuous use. It could be argued, therefore, that the end-of-life scenario is irrelevant in assessing the environmental profile of a chair and certainly of less importance than some of the other factors considered. Conversely, the choice of material has the single biggest bearing on the environmental impact of a plastic chair; therefore, it could be argued that the score awarded for this criterion could even be given more weight. Weighting of scores could impact overall rankings and this should be considered when interpreting results.

4. Results

Overall ERPR scores were calculated by summing the scores awarded to each of the five factors (materials, weight, transport, end-of-life and appeal), resulting in a total score out of twenty.

Out of the sixty designs evaluated, one achieved an excellent rating, with seven more achieving significantly above-average scores (Table 4.2). Rather than focus on individual scores my primary interest is to examine the designs which achieved the highest ERPR scores. By analysing these top-scoring designs, valuable insights can be gained for those interested in working with renewable plastics and developing plastic products with lesser environmental impacts.

Unsurprisingly, seven of the eight best-scoring designs were monoblocs, reflecting the efficiency of the one-shot injection-moulding process. Overall, the thirty-three monoblocs scored an average of ten points out of twenty, compared with the eight points for the remaining chairs. The highest scoring, *Bell Chair*, is manufactured in less than one minute using waste PP recycled waste from local industrial facilities and emerges from the production line stacked vertically, twelve units high, ready to be shipped (Figure 4.7). Weighting in at just 2.7 kilograms the *Bell Chair* was the lightest design included in my study and sets a new benchmark for dematerialization efforts. The manufacturer claims the chair can be recycled at the end of its useful life completing a perfect score for four of the five attributes survey. Although the *Bell Chair* is priced extremely competitively, only one variant of the design is offered in three colours (since increased to four), which impacted the scores awarded for appeal.

Out of the top-performing designs, only one chair, the *Hemp Chair*, was made from renewable organic materials (see Figure 4.10). As per predictions made by the waste hierarchy analysis, I had anticipated that designs made from renewable organic sources would achieve superior ratings due to their avoidance of fossil

Table 4.2 Overall Environmentally Responsible Product Rating (ERPR) scores

Score	Rating
17–20 (n=1)	Excellent
13–16 (n=7)	Above average
9–12 (n=27)	Average
5–8 (n=21)	Below average
1–4 (n=4)	Poor

n=number of chair designs allocated to each category

EPRP – highest-scoring designs 2007–2022

Figure 4.7 *Bell Chair*, Konstantin Grcic for Magis, 2020. Made from recycled post-industrial PP collected from waste created within the same production facilities and other local manufacturers. Chairs take less than one minute to mould and emerge from the production line stacked vertically twelve units high ready to ship. Manufacturer Magis claim the chair can be recycled at the end-of-life. Courtesy Magis.

Figure 4.8 *Remind R*, Eugeni Quitllet for Pedrali, 2021. Made from 50 per cent post-consumer waste and 50 per cent post-industrial waste PP. Chairs can be stacked four units high for shipping. Only available in one colour. Manufacturer Pedrali claim the chair can be recycled at end-of-life. Courtesy Pedrali.

Figure 4.9 *Sissi Chair*, Tramontina & Braskem, 2019. Made from recycled PP. Weighing just 2.8 kg the Sissi is available in only two colours with an armchair version also available. The *Sissi* is offered at a very low price. Courtesy Tramontina.

Figure 4.10 *Hemp Chair*, Werner Aisslinger, for Moroso, 2011. Made from natural fibres bonded with water-based glue. Extremely light chair that can be stacked for transportation (at least four units high) and is available in a variety of bright colours. Courtesy Alessandro Paderni for Moroso.

Figure 4.11 *Sai Go Green*, S.CAB, 2020. Made from 60 per cent recycled PP using renewable energy. Chair can be stacked six units high for shipping. Only two colours are offered. Manufacturer S.CAB claim the chair can be recycled at end-of-life.
Courtesy Studio Odeon for S.CAB.

Figure 4.12 *Ginevra Go Green*, S.CAB, 2020. Made from 60 per cent recycled PP using renewable energy. Chair can be stacked six units high for shipping. Only two colours are offered. Manufacturer S.CAB claim the chair can be recycled at end-of-life. Courtesy Studio Odeon for S.CAB.

Figure 4.13 *Hug Go Green*, S.CAB, 2022. Made from 60 per cent recycled PP using renewable energy. Chair can be stacked six units high for shipping. Seven colours are available, but price is more expensive than the other two S.CAB designs featured above. Manufacturer S.CAB claim the chair can be recycled at end-of-life. Courtesy Studio Odeon for S.CAB.

Figure 4.14 *Green Chair*, Javier Mariscal for Mobles114, 2016. Shell is made from recycled PP with wood and metal bases available. A range of four colours is offered. Chairs are transported in cartons of two and must be disassembled to facilitate recycling. Courtesy Mobles114.

fuels. However, scores for those chairs were in line with or below average, often constrained by the fact that material weaknesses (particularly of bioplastics) were compensated for by creating bulky, relatively heavy designs which cannot be stacked or dismantled to achieve transportation efficiencies and require specialist processing at the end-of-life (Figures 4.15 and 4.16).

Four of the eight top-scoring designs were awarded the full four marks for materials as they are monoblocs, constructed entirely from either a single material that is recycled (at least 90 per cent) or from renewable organic sources (see Table 4.3). The chairs from S.CAB (*Sai* [Figure 4.11], *Ginevra* [Figure 4.12] and *Hug* [Figure 4.13]) are made from 60 per cent certified recycled plastic (a mix of post-consumer and industrial waste). Despite these mixed feedstocks the manufacturer claims that these chairs are recyclable leading to the overall high scores awarded to these designs. It should also be highlighted that the plastic used to manufacture these three designs was recycled using renewable energy, further improving the environmental credentials of these chairs. The

Examples of chairs made from renewable organic sources

Figure 4.15 *The Coastal Furniture*, Nikolaj Carlsen, 2019. Made from seagrass with recycled wooden legs the chair weighs over 8 kg and is only available in one variant and one colour. Courtesy Nikolaj Carlsen.

Figure 4.16 *Albert Kuip Coffee Chair*, APE for Zuiver, 2021. Made from 42.5 per cent coffee waste combined with PP on an ash base. The chair weighs 4 kg but cannot stack and cannot be recycled at end-of-life. Only one variant is offered with one colour choice. Courtesy Zuiver.

Table 4.3 Detailed ERPR scores achieved by the best-performing chairs

Chair	Material	Weight	Transport	Appeal	End-of-life	ERPR
Bell Chair	4	4	4	2	4	18
Remind recycled	4	3	3	1	4	15
Sissi Chair	4	4	0	2	4	14
Hemp Chair	4	4	3	1	2	14
Sai Go Green	1	3	4	2	4	14
Ginevra Go Green	1	2	4	2	4	13
Hug GO Green	1	2	4	2	4	13
Green (wood leg version)	3	2	3	2	3	13

A score of zero to four points was awarded for each attribute giving a maximum potential ERPR score of twenty points. Note: scores are based on information provided by the manufacturers which has not been independently verified.

Green Chair designed by Javier Mariscal for Mobles114 is noteworthy as it is the only design that is not a monobloc yet still achieved an above-average rating (Figure 4.14). The *Green Chair* is shipped in cartons of two units which need to be assembled by the end-user, boosting the score awarded for transportation.

Six of the seven chairs awarded the highest ERPRs had all been designed to stack or pack in cartons containing at least four units during transport, resulting in above-average scores for this attribute. The *Sissi Chair* is also designed to stack up to eight units high, but details of how the product is shipped were unavailable, therefore it is assumed that chairs are shipped individually (Figure 4.9). All the high-scoring designs made using recycled plastics achieved top scores for their end-of-life prospects, as they are all designed to be recycled using widely available (mechanical recycling) infrastructure (with only the *Green Chair* requiring disassembly prior to processing).

Three of the top-scoring monoblocs weigh less than 3 kilograms, placing them in the highest-scoring category for this attribute (injection-moulded monoblocs weighed 4.2 kg on average, with all sixty chairs averaging 6 kg). All the top-scoring designs achieved relatively weak scores for appeal, due to limited colour choices and lack of variants (selection of base and shell options). While those compromises are often made to contain costs and reduce waste, the lack of choice offered to purchasers is almost certain to negatively impact sales potential.[3] The three dimensions of appeal measured in this analysis do conflict – offering a wide range of colours and variants to broaden the appeal of a product can only increase costs, which must be passed on through higher retail prices. When developing chairs for the mass market, choice is often sacrificed in the trade-off against lower prices. Overall, only two of the sixty designs included in the analysis achieved an above-average appeal rating of three points, with none achieving the highest score.

The overall findings demonstrate that it is possible to achieve good environmental ratings without experimenting with new renewable organic-based materials. Lightweight stackable chairs made from recycled plastic can achieve high ERPRs. The study also revealed that access to production methods requiring high up-front investment, such as injection moulding, is also not a prerequisite to produce environmentally responsible plastic chairs. The *Hemp Chair*, which achieved a top rating, is made using thermoforming, a more accessible and cost-effective method that avoids the need for expensive moulds, demonstrating that sustainable designs can be created without access to advanced injection-moulding facilities. A further five designs that achieved ERPR scores in the average category were made using compression moulding or thermoforming techniques, further demonstrating that environmentally responsible design can be achieved without experimenting with bleeding edge materials or state-of-the-art production facilities.

It is important to note that the results from this analysis should be approached with caution, as the information used was provided by the manufacturers and has not been independently verified. For example, the manufacturer of the highest-ranking design claims that chairs can be recycled at the end of their useful life. However, fibreglass has been added to reinforce the recycled PP and the chair is not embossed with a resin identification code (RIC), making it unlikely that a discarded chair would be accepted by a mechanical recycling facility. Additionally, with no detailed information available on resources consumed during production or the origin of outsourced components, the weight of a product has been used as a surrogate to compare resource and energy efficiency. The environmental benefits of thermoforming, which generally requires less temperature and pressure than injection moulding, were therefore not acknowledged in this analysis. Similarly, the tool is not sufficiently sophisticated to reward manufacturers who are using renewable energy or engaging suppliers that do. Different feedstocks and material manufacturing processes cause different environmental impacts – recycled ocean plastic and plastic recycled from the waste of an industrial process using virgin plastics are likely to cause different environmental impacts and the ERPR tool does not attempt to quantify these differences. Consequently, these limitations can be expected to impact the overall ranking of environmental performance.

Moreover, this analysis, like cradle-to-gate LCAs, fails to consider the projected lifespan of the products. High-quality, well-crafted chairs can be expected to last longer, postponing any end-of-life impacts, with carbon components storing carbon and preventing it from entering our environment. While manufacturers can make claims about the anticipated lifespan of their products, such claims are challenging to evaluate and are not considered in this analysis. Also note that this and other eco-audit tools focus on the economic and environmental credentials of products but fail to consider the social dimension of sustainability.

This simplified ERPR tool enables comparisons between designs within the same product category using only publicly available data, which is the typically limited information available to prospective purchasers from manufacturers' websites and retail point-of-sale material. ERPR is a highly simplified and modified version of the original tool developed by Graedel and is not intended to compete with more sophisticated eco-tools that evaluate the environmental credentials of products. ERPR can serve as a helpful guide for design decisions, in contrast to most tools that only become useful when accurate data is available, often after a design enters production.

5. Summary

Consumers and organizations that prioritize environmental considerations in their decision-making must assess ecological impacts based on the information available. Existing eco-audit tools, designed to enable an accurate assessment of the environmental impacts inflicted by products, require highly detailed information, which was regrettably not available for this study, or to prospective end-users. Consequently, I developed a simplified methodology with the overarching aim of identifying the ecological ramifications inherent in the use of renewable plastics in product design. The ERPR tool facilitates such a comparison and was used to assess sixty chairs made with renewable plastics. Despite its limitations, the ERPR tool serves as a rudimentary but practical resource for comparing the environmental impacts of products within the same category and identification of those products with lighter environmental footprints (see Appendix in this book for more details).

For each chair the ERPR tool was used to award scores for five factors: materials, weight, transport, end-of-life and appeal, with a theoretical maximum score of twenty points attainable. One design achieved an excellent rating, while seven scored significantly above average. Results were then used to examine the designs that achieved high environmental ratings. Surprisingly, only one top-scoring design, the *Hemp Chair*, was made from renewable organic material. Designs made from bioplastics tended to be bulkier, in efforts to compensate for the reduced mechanical properties of these materials. The remaining top-scoring designs were injection moulded from recycled plastics. However, access to expensive injection-moulding technology is not a prerequisite, with thermoforming used to develop a top-rated design. All the top-rated designs received low scores when evaluated for appeal due to the limited number of colours and variants offered, which could negatively impact sales potential.

Results derived from this analysis are of little direct relevance beyond the sphere of those engaged with the production and consumption of plastic chairs. The next chapter shifts the focus towards the broader role of design in the creation of consumer goods. Results from the ERPR analysis are further interrogated and combined with insights gleaned from interviews with product designers experienced with renewable plastics. I identify common attributes that can be extrapolated and applied more broadly by those working with plastics to develop consumer goods.

5   The agency of design

Designers are traditionally primarily tasked with responsibility for shaping the form and aesthetic aspects of a project. Their expertise and experience are, however, valuable assets that can assist other actors with decision-making across the entire process of creating and selling a product or service. Those decisions often affect the environmental performance of a design. While the transition from virgin plastics to renewable carbon-based materials is expected to deliver emission reductions, environmental benefits are optimized only if the sustainability implications of all facets of a design project across its entire life cycle are considered as illustrated in the previous chapter.

This chapter examines the pivotal role that design can play in influencing major decisions that affect a plastic product's environmental impact from its inception, through production and beyond the factory gate. Lessons gleaned from the past projects which have already succeed or failed in delivering significant environmental benefits can be applied to the planning and development of future designs.

Despite compelling pressure to continue business as usual, the use of virgin plastics will face escalating scrutiny, as the health and environmental impacts of the material become more increasingly evident. Renewable plastics are poised to gain greater favour with manufacturers as demand from purchasers (including consumers and business purchase decision makers) increases.

1. What agency does design hold?

In 1971, Victor Papanek accused designers of being responsible for planned obsolescence and choosing materials and processes that result in pollution. He concluded his argument with the often-cited quote, 'there are professions more harmful than industrial design, but only a few of them' (V. J. Papanek, 1971, p. ix). Papanek's views were not well received when first published, he was derided and verbally attacked by his peers who then forced him to resign from the industry association (Andrews, 2015, p. 311). Papanek, however, succeeded in igniting a long-running debate about the ethical, social and moral responsibilities of designers.

Can environmentally conscious product designers be expected to have any agency in how their designs are manufactured and especially in the choice of materials? In more recent times, some theorists are calling for extreme action by product designers – who they say have the moral responsibility to take the leading role in guiding society toward a more sustainable existence (Fry, 2009; Irwin & Kossoff, 2017; Tonkinwise et al., 2015). Tony Fry encourages designers to end their complicity with the creation of defuturing conditions and 'place the needs of the market in second place to the political-ethical project of gaining sustainability' (Fry, 2009, p. 46).[1] Fry challenges designers to ask themselves what they 'can see taking the future away, and what can [they] do to reduce the impacts of [their] own actions which defuture?' (Fry, 2020, p. 143).

But do product designers have the agency needed to meet these lofty demands? Dutch designer and academic Jan Joore emphasizes the significance of the size and complexity of a project. Where a single problem owner or client commissions a project they ultimately decide the course of action to be taken. As the scope of systems get larger, the precise agency of any actor, including designers, becomes less clear (Joore, 2010, p. 197). A designer working as part of a large team on components for a consumer product has less agency than when specifically commissioned to design a standalone item. Dutch designer Bertjan Pot (b.1975) echoed this sentiment, commenting during an interview with me that 'if you want to change something in a car, imagine how many meetings there will be to make a small change?' The complexity increases further for those designing transportation systems (Joore, 2010, p. 197). Indeed, my research found that when a designer acted as a maker, they often retained control over design decisions. This observation held true both when designers had direct responsibility for production through their own studios (such as Tom Fereday and Louis Durot) and when they contracted out manufacturing to their own specifications (for instance, DesignByThem and Bertjan Pot).

My interviews with designers for this study revealed a wide range of attitudes towards their influence of sustainability considerations on their overall design process. A small number of designers freely admitted that sustainability is not their primary consideration when creating new products. Those designers included French industrial designer Louis Durot (b.1939) and Ron Arad (b.1951) who reported:

> When you work you have to look at the effect of the process on the environment. It is not my first consideration; I am not expecting brownie points

At the other end of the spectrum, some designers expressed extreme concern over the sustainability impacts of their work, to the extent that several reported considering leaving the profession, with Ander Lizaso (Spanish industrial designer) among them:

> Actually, in our studio, we are very critical about the role of the designer and design in general. If you're really deep and honest about it, the best thing we could do is to just stop ... I mean the amount of pollution that we create ... But if we are really going to not harm, we shouldn't be a [cog] in this assembly of consumption and production.

Dan Armstrong, from Australian design agency Formswell, also discussed how sustainability issues had made him question his agency's work:

> How much do you go back and indulge your beautiful plastic chair and keep making those things? When do you start becoming guilty about everything that you're doing? It is a weird line there ... Do I keep going? Do I adjust what I'm doing?

In response to those concerns and a recognition of the environmental impacts caused by their work, designers reported employing various strategies in efforts to lessen those effects. Focusing on developing lasting, quality products, likely to last a long time thereby prolonging any end-of-life impacts was frequently nominated as a valuable contribution toward sustainability. Edward Barber and Jay Osgerby (for UK-based industrial design studio Barber Osgerby) are among those who expressed this, claiming to only work with manufacturers that make high-quality products, designed to last. John Tree, senior designer at Japer Morrison's office echoed similar views, reporting:

> Every designer is very aware of the impact of making more stuff has on the world really. There's a focus in our office on working with brands

that believe in longevity and sustainability and things that are designed to function rather than fashion or something like that. I think that Jasper's products are a testament to that in a way, a lot of the things that he designed maybe 20 or 30 years ago are still in production now, still selling very well and it's a very deliberate part of this office's work is to try to find, to treat the object as the client and not the time it is designed in.

Those claims are borne out by a survey of furniture manufacturers undertaken by London-based design practice, Dodds & Shute (Dodds & Shute, 2020). Frustrated by the lack of readily available information to verify sustainability claims and only wanting to use furniture produced by companies committed to sustainability, Dodds & Shute conducted a survey of their suppliers to determine their sustainable practices. In addition to identifying the manufacturers with the most demonstrable commitments to sustainability, Dodds & Shute went on to identify the designers who had completed the most projects with these organizations. Barber Osgerby, together with Jasper Morrison, ranked among the top six designers (out of five hundred) based on having completed the largest number of projects with manufacturers who demonstrated the most commitment to sustainability (Dodds & Shute, 2020, p. 20).

Although not included in the Dodds & Shute survey, Gregg Buchbinder, the owner, and CEO of American manufacturer Emeco, also prioritizes his attention to ensure product longevity. He believes the best way to improve the impact of a product is to reduce the energy needed to manufacture and design the product to last as long as possible. He claimed: 'our main environmental focus is to create chairs that last – we focus on lengthening the life first'.

Lizaso reported that his design studio attempts to make a difference by only working with organizations with a committed to sustainability principles. However, he highlighted that it is easier for designers with a significant public profile to make sustainability commitments, but much harder for those without an established name. Lizaso also reported struggling with the conflict of designing more furniture while being increasingly concerned with sustainability and over-consumption.

Designers are not the only actors influencing the sustainability of a design. Buchbinder outlined the important role of manufacturers in driving sustainability:

With regards to the role of product designers, their role—just like ours as a manufacturer—is evolving to reflect the now urgent need to address our environmental impact. Every product development process must take into consideration the complete product lifecycle and give careful consideration to both material and process. This is the joint responsibility of product designers and manufacturers. But we believe that it remains the responsibility of any manufacturer to explore new, sustainable solutions for their production. Our experience is that product designers welcome the opportunity to experiment with new, sustainable materials. Unfortunately, not all manufacturers offer that possibility.

These comments highlight the challenge faced by designers with a commitment to sustainability. They must secure support from manufacturers who hold similar opinions and have access to the resources and expertise needed to develop their designs.

Despite these obstacles, designers can and should play a leading role in initiating and encouraging the use of renewable plastics. Philippe Starck, renowned for his commercial success career thanks, in part, to his work in plastics, including his numerous chair designs for Kartell, told me about the responsibility he takes in such matters:

> Before [in the 1980s] this plastic bashing was stupid and this eco trend was stupid. But now it's serious, we have problem, we have to solve it … The eco trend today is not a trend it is something vital … It's not a choice, we have to work with it. A producer like me has to take their own responsibility. And I do it, I take responsibility, especially on plastic.

It is no longer sufficient for designers to encourage their leaders and educators to declare a climate emergency while continuing with business as usual. Designs destined to be manufactured using fossil derived materials and/or processed using fossil-based energy must be challenged as suitable renewable carbon-based alternatives become available. Designers can use their agency to influence and drive the uptake of renewable plastics.

Figure 5.1 highlights the decisions most likely to be influenced by the designer. While the designer can be expected to exert the most control over the four topics shown in the centre of the diagram – aesthetics, weight, form and transport – they can only contribute, to varying degrees, to the other decisions highlighted. For example, cost of production is ultimately the responsibility of the furniture manufacturer, but design decisions directly impact those costs. Many decisions are ultimately made by the manufacturer, but it is likely that the designer and in some cases the material manufacturer will be consulted for their expert advice

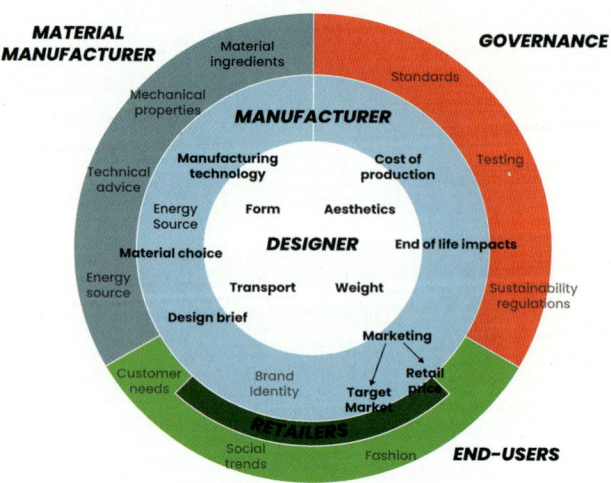

Figure 5.1 Chair design – sphere of influence. The decisions that need to be taken when developing a design are shown with those influenced by the designer highlighted in bold. Where decisions are impacted by other actors (labelled in dark blue caps) this is indicated by the positioning of the decision labels.

or opinion. The agency of the designer in the decisions process is examined in this chapter beginning with those decisions where potential impact is most pronouncedly.

a. Aesthetics

In Chapter 4, I highlighted that a product only actually delivers an environmental benefit if it is purchased at the expense of a less-sustainable product. Even products boasting the highest sustainability credentials are simply a waste of resources if they fail to sell – a fact that is often overlooked when evaluating eco-credentials. Making renewable plastics appealing to end-users is an important task for the designer. A potential end-user is likely to form their first impressions of a product based on its form and appearance, which could be to the detriment of products made from renewable plastics. Glossy surfaces with a shiny or lustrous, metallic appearance are appealing to many observers. The importance of the surface appearance of virgin plastic is well recognized by the industry, which focuses substantial efforts toward its refinement.[2] Recycled plastics are often less shiny and cannot reach the same consistency and homogeneity as their virgin fossil equivalents.

Surfaces can be slightly rough to the touch, with swirls or other blemishes on the surface. When presented to a market grown accustomed to the perfection achieved by virgin polymers, the imperfect finish of recycled plastics can deter buyers, as Emeco discovered to their cost. CEO, Gregg Buchbinder reported a buyer rejecting an entire shipment of *111 Navy Chairs* (made from recycled PET) on arrival in Japan, as they found the level of finish unacceptable.[3]

Technological development has driven an expectation of perfection, with quality control procedures fine-tuned to eliminate any deficiencies resulting in products with a perfectly homogenized appearance. However, the inherent heterogeneity of recycled plastics presents opportunity, as imperfections can be positioned as enhancing the value of a product, and can even be considered endearing (Karana et al., 2017, p. 209). The difference in appearance between recycled and virgin plastics can be exploited to signal the ecological credentials of the material. However, while a product displaying a recycled look might appeal to some, there is a danger that the mass market will judge the telling signs of human detritus as austere, inferior or cheap and reject it. The challenge for designers working with recycled materials is to embrace the natural colour shades, random patterns, rough textures, imperfections, irregular surface characteristics of the materials and shift cultural value systems to render these qualities appealing to end-users. Renewable plastics create unique opportunities for designers to push beyond imitating natural materials and foster the development of new sensual experiences.

Philippe Starck was among the first designers to take on the challenge of exploiting the inherent aesthetic properties of recycled plastics for a chair. During our conversation, he claimed that far from being concerned by this challenge he relished the opportunity to create new sensual experiences as he was bored with working with 'perfect plastic'. The recycled polymer mixed with wood waste used for the *Broom Stacking Chair* created a slightly rough texture and a matt finish, unique to the market when it first appeared in 2012 (Figure 5.2). Starck told me that he 'loved the idea to have a new material, natural with life and defects'.

Renewable plastics are often only available with limited colour choices, particularly where post-consumer feedstocks are used. Thomas Pedersen used recycled household plastics to produce the shell for the *Falk* series made by Houe and reported, 'The only problem is that you can't get it in every colour. You can only get it in black.' Where more colour choices are available, these are often restricted to limited earthy, muted pallets. Those muted pallets can be appealing to those consumers wishing to make a public display of their environmental credentials but risks limiting products to a small market niche. Some designers, including

Figure 5.2 Detail showing textured surface of *Broom Stacking Chair*, designed by Philippe Starck for Emeco, 2012. Courtesy Emeco.

Pedersen, have addressed this deficiency by offering a range of colourful (recycled polyester) padded seats as optional extras, likely broadening appeal while potentially improving profitability (see Figure 4.6).

Bioplastics are also often only available in a limited palette of colours that might be described as 'mushy and muddy' as German industrial designer Konstantin Grcic claimed in our interview. Some designers have addressed that issue by performing their own experiments to broaden the range of choices available. For example, Spanish designer, Ander Lizaso reported that he had developed his own colours for the bioplastic shells of his *Kuskoa Bi* chair (Figure 5.12). The designer emphasized to me the role of trial and error and experimentation to achieve the final range of colours.

Navigating the confines of the aesthetic limitations of renewable plastics presents designers with both their main challenge and the best opportunity to effect change by improving the appeal of their products. Different strategies have already been developed by designers working with renewable plastics to overcome any perceived sensory deficiencies. These strategies can be categorized as: avoid, embrace or celebrate.

Some designers have avoided the issue altogether by using renewable plastics that look virtually identical to fossil plastics. Emeco invested in improving the appearance of recycled PET to make it indistinguishable from virgin material when used for its *111 Navy Chair* (Figure 5.3). DesignByThem explored a far less expensive solution for their *Butter* series, folding HDPE panels made from a central core of recycled HDPE (80 per cent) sandwiched between veneers of virgin HDPE (Figure 5.4). This hybridized material allows some of the properties of the virgin material (including those that affect the senses) to be retained without impacting end-of-life prospects. The design is offered in a range of bright primary colours, while the surface displays the perfect homogeneous finish typically associated with virgin polymer (Figure 5.5). A potential benefit of this approach is that it avoids the stereotypical 'green' look often associated with recycled plastics, which may limit appeal to a relatively small market segment.[4]

Materials can be concealed in other ways; Bosch are experimenting with using recycled plastic for the base structure of a dishwasher, a component that will never be seen by end-users (Laird, 2022). Just replacing this single component

Designs that avoid the look of recycled plastics

Figure 5.3 *111 Navy Chair*, Emeco, 2010. An update on Emeco's original Navy Chair made from recycled aluminium and designed in 1944. This more recent version is made from recycled Coke bottles, originally 111 were used but this has increased to 170, with the proportion of fibreglass decreasing from 35 per cent to 10 per cent. Eleven colours are available. Courtesy Emeco.

Figure 5.4 *Butter Seat*, Nicholas Karlovasitis & Sarah Gibson for DesignByThem, 2011. Made from 80 per cent recycled HDPE with a veneer of virgin feedstock which allows the chair to be offered in a range of nine bright colours. Chair arrives flat-packed from the manufacturer and is assembled locally for distribution. Courtesy Pete Daly for Design ByThem.

Figure 5.5 *Butter Seat* from DesignByThem is made from recycled HDPE boards veneered with virgin HDPE, allowing a range of bright colours to be offered. Courtesy Pete Daly for Design By Them.

could result in saving 2,500 tons of crude oil annually according to Bosch. This initiative hints at the enormous potential for renewable plastics to be used for components within consumer goods. Numerous plastic components required for consumer goods remain concealed from end-users throughout their functional lifespan, rendering them ideal candidates for redesigning with renewable plastics, regardless of whether their appearance is considered too unattractive to be visible.

Emotions are not evoked by visual stimuli alone. Barber Osgerby's *On & On Stacking Chair* for Emeco (Figure 5.24), like Starck's *Broom Stacking Chair*

(Figure 5.18), is made from recycled plastic mixed with wood. The material developed for Emeco was designed to provide adequate strength while minimizing the use of fibreglass. However, the process imparts a unique textual experience to the finished product and makes it warmer to the touch. Features that Jasper Morrison noticed when he used the same material for his *1 Inch Reclaimed* (Figures 5.6) series (also for Emeco) rating the material as having a 'warmer touch, more natural touch than raw polypropylene' (Designboom, 2018). In addition to this unique sensorial experience, the use of wood – a familiar raw material – might also signal to users that the product to be more 'environmentally friendly' or 'natural', encouraging appreciation for the product and possibly even extending its useful life (Karana et al., 2017, p. 209).

The *Odger* for IKEA also proudly displays a rough textured finish due to the addition of wood to the recycled PP (Figure 5.7). Colour variations of recycled plastics are often regarded as a substantial deterrent by designers and manufacturers alike. But some have chosen to celebrate this property, even promoting it as a benefit. NCP hails the irregular patterns that appear on the surface of its *S-1500* chairs (Figure 5.8), claiming them as marble patterned 'celebrating the ocean's movements' and creating a unique finish for every individual chair (NCP, n.d.).

Designs that embrace the look of recycled plastics

Figure 5.6 *1 Inch Reclaimed*, Jasper Morisson for Emeco, 2018. The original version was made from 1-inch square aluminium tubing referencing Emeco's original *Navy Chair*. This monobloc version is made from 88 per cent waste PP with 2 per cent waste wood fibre and 10 per cent fibre glass. Courtesy Emeco.

Figure 5.7 *Odger*, John Löfgren & Jonas Pettersson for IKEA, 2017. Detail showing the rough texture featured on the Odger designed for IKEA. The chair is made from 70 per cent PP, of which 55 per cent is recycled, and 30 per cent wood. Courtesy Jonas Lindstrom for Form Us with Love.

Designs that celebrate the look of recycled plastics

Figure 5.8 *S-1500*, Snøhetta for NCP, 2019. Detail showing the marbleized appearance created by using recycled fishing nets to create the shell. Courtesy Bjornar Ovrebo for NCP & Snøhetta.

Grcic even went as far as to 'fake' the recycled look by adding 'black sprinkles' to the material used to make the *Bell Chair* (Figure 5.9). Through experimentation the designer discovered that the recycled PP included some impurities which would display on the surface of the chair. Concerned that end-users might consider this a mistake or fault, additional impurities were deliberately added to the material to give a more uniform appearance. Grcic claimed: 'it tells the story of the recycled material and makes the look and feel different to the other plastic'. By celebrating the imperfection of the material, the aura of unique handcrafted design has been introduced to mass-produced products. The opportunity to own a unique design in the age of mass production presents an appealing proposition, as witnessed by the rise of craft movements in high-income countries in the twenty-first century.

Figure 5.9 *Bell Chair*, Konstantin Grcic for Magis, 2020. Detail showing the speckled surface which highlight the use of recycled PP. Courtesy Magis.

Bioplastics offer the potential to be fine-tuned – their characteristics, including appearance, can be adjusted to suit the task at hand and the taste of the designer. In our interview, Mayda Diaz, from the Australian bioplastic supplier Bambacore, explained that by adjusting the proportion of bamboo fibres include in their bioplastic range, both the look and mechanical properties of the material can be controlled. Impact resistance can be improved by increasing the fibre loading and adding modifiers, for example. This technique can also be used to modify the surface, with mirror, polished, glossy, matt, smooth and rough textured versions available. Elvin Karana, a professor of materials innovation and design, contends that the unique properties of bioplastics present opportunities for designers to explore all aspects of the material including touch, smell and appearance (Karana et al., 2013, p. 203). Designers working with flax and hemp proudly display the use of natural fibres on surfaces, imbuing each chair with a distinctive and unique finish while encouraging an appreciation of their more neutral tones (Figure 5.10).

Figure 5.10 *Jin Chair*, Jin Kuramoto for Offecct, 2018. The use of flax is clearly visible on the surface of the design. Courtesy Thomas Harrysson for Offecct.

The imperfect nature of renewable plastics might also extend the longevity of products, as they often wear the effects of time more gracefully than the perfect shiny, smooth surfaces of their virgin counterparts. Imperfections can be interpreted as more natural, helping to create an emotional bond with the user. Traditional materials gain a patina over time which is often interpreted as a sign of maturity and can be highly valued, extending product longevity (Karana et al., 2017, p. 212). Indeed, in 1995, Victor Papanek predicted that environmentally and socially orientated design of the twenty-first century would include 'graceful aging' as the first fundamental principle, as materials that aged well hold great appeal (V. J. Papanek, 1995, p. 24).

Renewable plastics create the opportunity for designers to completely reimagine our relationship with the material and reshape our cultural norms. Modifying market expectations away from pristine surfaces by promoting an appreciation of 'graceful aging' with signs of use and wear adding to the history of the material. The challenge is to find connections between the unique properties of the material, production processes and establish satisfying emotional connections with the viewer, purchaser and end-user (Cleminshaw, 1989, p. 130). Responding to this challenge Lievore + Altherr Désile Park created an organic texture based on concentric tree trunk rings for the *Adell* collection (Figure 5.11).

> The texture evokes natural materials without being a representation – accepting and celebrating the unregular, the unperfect. Scratches to the surface dissolve in the texture, acting like a patina, not wear.
>
> (Altherr, 2020)

Growing and securing consumer acceptance of the often-unique aesthetic attributes of renewable plastics in consumer products is essential for the niche to develop. The different strategies explored by these groundbreaking designers provide a preview of the new horizons awaiting further exploration by those choosing to work with renewable plastics.

Figure 5.11 *Adell*, Lievore + Altherr Désile Park for Arper, 2020. Shell made from 80 per cent recycled PP. Courtesy Arper.

b. Weight

For plastic components in products that do not consume resources during the operational phase, by far the most substantial environmental impact is caused by the manufacture of the plastic itself. An environmentally conscious designer's priority should be to minimize the quantity of material used in any of their plastic creations. Lightweighting or dematerializing plastic products not only reduces the energy consumed during their manufacture but delivers downstream benefits, through savings in both transportation costs and reduced end-of-life impacts (i.e. less to recycle or landfill). Dematerialization efforts can be expected to be actively supported by manufacturers, as reducing the amount of material and energy required to manufacture a product helps to maximize profits.

However, there are finite limits to the savings that dematerialization can deliver. As we have seen, industry has focused on dematerialization as a strategy to improve profitability while delivering environmental benefits, often coincidentally. Two decades after Starck developed his 3.5 kilograms *La Marie* for Kartell, the lightest chair available weighs 2.7 kilograms. While this represents an impressive 23 per cent reduction, the absolute saving of 0.8 kilograms is small compared with the gains made during the two decade prior to *La Marie* (around 2 kg). Further reductions are likely to be smaller again, as other design specifications for strength and stability crucial to the satisfactory performance of the chair must take precedence over weight-saving objectives (Elzen, 2004, p. 1). Despite these limitations, advances in materials and manufacturing processes continue to create opportunities to further optimize designs, and these opportunities should be fully exploited.

c. Form

Manufacturers often favour producing timeless or classic designs. By deliberately avoiding transient stylistic trends, the life of products can be extended both in production and in the use phases. Countless plastic chairs have been created over the past eighty years but only a few have withstood the test of time, continuing to stimulate sufficient demand to keep them in (sometimes continuous) production for decades. Perhaps more importantly, many examples of those designs remain in constant use, sometimes even becoming heirlooms, postponing end-of-life environmental impacts. Those design classics, with their proven timeless stylistic allure, add to the bank of ideas providing inspiration for others. By revisiting tried and tested solutions designers minimize the risk of commercial failure. Spanish designer Eugeni Quitllet referred to this during our interview: 'furniture designers are inspired by the interpretations of other designers who successfully made use of technological possibilities in good products'.

Revisiting classic designs from the recent past can also help to create an air of familiarity, potentially connecting prospective purchasers with nostalgic memories and promoting an emotional bond with the product. Designs that have endured are proven to be fit for purpose and aesthetically durable. In the case of chairs, prospective purchasers can be reassured the design has withstood the ultimate test of prolonged real-life use and is considered comfortable and fit for its task.

In Chapter 1, I included a quote from Elvin Karana who advised designers to stick with a familiar material if they were exploring novel shapes (Karana et al., 2013, p. 6). Many of the designs in my study heed this advice, sticking to familiar shapes while experimenting with novel materials. Ironically, the mid-century period (when many designers were experimenting with shape and material at the same time) is the preferred source of inspiration for many of these designs. By deliberately minimizing ornamentation that could fall victim to transient fashion, leading modernist designers successfully developed products, including chairs, which continue to sell while inspiring new generations of designers.

Robin Day's *Polyside* armchair inspired Lizaso (and his then partner Iratzoki) when developing the *Kuskoa Bi*, the profiles of the shells of the two chairs are similar (Figures 5.12 and 5.13). Mater revised a Jørgen & Nanna Ditzel design from 1955 to develop their *Ocean Chair*, only increasing the size of the chair by

Figure 5.12 *Kuskoa Bi*, Ander Lizaso and Jean Louis Iratzoki for Alki, 2015. Courtesy Mito for Alki.

Figure 5.13 *Polyside Armchair*, Robin Day for Hille, 1967. Courtesy Hille.

5 per cent, to accommodate an increase in the average weight of people since the chair made its first appearance in 1955 (Figure 5.14). The *S-1500* is a 'structural redesign' of Bendt Winge's classic R-48 chair from 1970, which has sold over 5 million units in Norway alone (Figure 5.15) (Matslinder, 2017).

Garcia's *Voxel Chair v1.0* (Figure 5.23) is an update of the cantilevered *Panton Chair*, available from 1967. Tom Price used an Eames armchair (from 1950) as a mould to make a direct impression of the chair in balls of PP rope (Figure 5.16).

While not inspired by a single work, references to previous design solutions can be found in other examples. The *Bell Chair* adopts L-shaped legs, long accepted as delivering the most economical use of material to provide the required strength (Figures 4.7 and 5.9). The shell of Laarman's *Puzzle Chair* (Figure 5.18) is similar to the Eameses'

Figure 5.14 *Ocean OC2 Chair*, Jørgen & Nanna Ditzel for Mater, 2018. Courtesy Mater.

Figure 5.15 *S-1500*, Snøhetta for NCP, 2019. A revision of the R-48 designed by Bendt Winge in 1970. Courtesy Bjornar Ovrebo for NCP & Snøhetta.

Figure 5.16 *Meltdown Chair*, Tom Price, 2007. A mould based on the Eames armchair is used to make an impression in a ball of nautical rope. Courtesy Tom Price.

Figure 5.17 *DSX Side Chair*, Charles and Ray Eames for Herman Miller, 1948. The more resilient H-shaped base replace the original X shape in 1954. Courtesy Vitra.

Figure 5.18 *Puzzle Chair*, Joris Laarman for Bits&Parts, 2014. The 3D files and printer settings are available for free on the internet. Individual puzzle pieces can be printed on a standard domestic 3D printer with broom handles used for legs. Courtesy National Gallery of Victoria, Melbourne.

Figure 5.19 *Polyside Chair*, Robin Day for Hille, 1963. Early use of PP to create the shell. Different bases were available to make the chair suitable for a range of tasks. In the 1970s, over thirty countries manufactured the chair which has sold millions of copies. Courtesy Hille.

Figure 5.20 *Flax Chair*, Christien Meindertsma for Label/Breed, 2015. Flax layers and PLA used to make a 660 × 100 cm sheet. A chair is cut and constructed from a single sheet. Courtesy Label/Breed.

plastic shell chairs (Figure 5.17), while Day's *Polyside Chair* (Figure 5.19) appears to be inspiration for the shell shape featured in Meindertsma's *Flax Chair* (Figure 5.20).

Other designers included in my study took a different approach and sought their inspiration from historical examples, or even the ancient forms found in the natural world. Many of the designs manufactured using new technologies such as 3D printing or AI were inspired by the natural environment. Designers working with new manufacturing technologies have gone beyond a study of form, concerning themselves with exploring the structural secrets developed by evolution of over billions of years. Nature has evolved to be conservative with mass, minimizing the amount of material needed to deliver the required function. Mimicking structure and process has allowed designers to explore new, potentially more efficient ways to create traditionally solid shapes.

Starck used Autodesk AI software, designed to mimic the growth of trees and bones, to minimize the quantity of material needed for Kartell's *AI Chair* (Figure 5.21). When we spoke, Starck acknowledged the similarities between nature-inspired Art Nouveau design, popular in the early years of the twentieth century, and his chair developed using the latest in software technology over a century later without inputting cultural, social or aesthetics references. Marcel Garcia is experimenting

Figure 5.21 *AI Chair*, Philippe Starck for Kartell, 2019. Artificial intelligence software was used to assist the designer in developing this chair. Courtesy Kartell.

Figure 5.22 *Voxel Chair v1.0*, Manuel Garcia for Nagami Design, 2016. Right – detail showing unique 3D printing process claimed to be based on nature and developed to minimize the impact of errors encountered during the 3D printing process. Courtesy © Nagami Design S.L.

with 3D printing techniques directly referencing structures in nature in efforts to minimize both material consumption and faults, which can arise during the printing process resulting in wastage (Figure 5.22). Rashid claims his bulbous *Siamese Chair* is 'amorphous, from the earth and an extension of nature' reflecting the characteristics of the bioplastic used for his design (Figure 0.3).

Form can also be affected, dictated even, by the mechanical properties of the particular material selected. There are risks working with new materials as mechanical limitations might only reveal themselves during the production process. For example, Rashid experienced issues working with bioplastics when developing the *Siamese Chair*; plans to develop a monobloc had to be aborted to accommodate the mechanical properties of the material. To succeed, design concepts were compromised to accommodate the technical limitation of the renewable carbon material. Rashid disclosed that the design has not been a commercial success, highlighting the financial and reputational risks designers expose themselves to when experimenting with untried materials.

d. Transport

When making decisions on form the implications for transport and storage must be considered. Many of the chairs I have studied (for example *AI Chair*, *Kuskoa Bi*, *Siamese Chair* and *Voxel Chair v1.0*) are not designed to stack. In many cases those chairs are packaged for transportation in cartons containing only one or two units, taking up space in containers and warehouses with negative financial and environmental implications. How a product will be transported is an important design consideration when creating a more sustainable design. When designing seating solutions transport can be optimized by creating stacking chairs but this does impose limits on the overall design, as the back legs of one chair usually need to slot through the shell or seat. However, if these design limitations are unacceptable there are other solutions to optimize for transport. For example, the *Odger* occupies less space when shipped, as the end-user is required to assemble the legs and shell (Figure 5.23). The *Butter Seat* arrives in Australia flat-packed from the Chinese production facility and is assembled for local distribution, helping to reduce the environmental impact of international transport (Figures 5.4 and 5.5). Experimentation with new manufacturing technologies can open new possibilities for innovative solutions to minimize transport impacts.

e. Material choice

With many actors involved in the development of a product, including clients, manufacturers, sales and marketing teams, ergonomics experts, brand consultants, material scientists and engineers, identifying the decision-making process for materials and defining the precise role of each actor is a complex process. Reporting on an experiment in the Netherlands with twenty Turkish professional designers, Elvin Karana et al. suggest that designers should shift from material selection to material inspiration. Rather than follow a traditional linear approach, for example,

Figure 5.23 *Odger*, John Löfgren & Jonas Pettersson for IKEA, 2017. *Odger* is delivered in four pieces designed for easy self-assembly, reducing transport and storage costs for client IKEA. Courtesy Jonas Lindstrom for Form Us with Love.

deciding to manufacture an object from plastic and then a class of plastic and then a particular brand, the designer should consider materials during the initial stages of design, allowing the material's properties to inspire the design (Karana et al., 2008, p. 1088). This approach suggests that designers have ultimate control over the selection process. The designers included in this study overwhelmingly reported that the decision on the specific type of plastic to be used is unlikely to rest with the designer alone for most projects. The increased complexity of manufacturing techniques has been accompanied by the development of specialists, who are employed or consulted by manufacturing organizations. It is these specialists that make recommendations for materials based on the technical requirements of the project, with the manufacturer often responsible for the final decision, discussed further in Chapter 7.

Reflecting the above findings, research participants who took responsibility for production through their own studios (e.g. Tom Fereday and Louis Durot) or by contracted out manufacturing to their own specifications (e.g. DesignByThem and Bertjan Pot) were more likely to have greater agency in the material selection process. Where designers provided their services to clients, creating designs for large-scale mass production, the influences of other actors became more prominent, especially as the client typically acts the role of manufacturer as well. Ron Arad illustrated this point by highlighting the difference between limited-edition studio pieces, where he maintains total control over the selection of materials, and high-volume production pieces for manufacturers, where the commissioning client has a 'lot of say in it'. Pot agreed, reporting that the manufacturers of his commissioned production pieces have the most agency in material selection.

Gabriel Chiave, studio manager for Marcel Wanders, suggested that designers do take the leading role in creating and problem solving, but claimed they are the least impactful in the final decision on materials and method of manufacturing. Designer and academic Trent Jansen highlighted that decisions around material choice can also be influenced less overtly:

> There's a commercial project that's being manufactured just recently, [they] have been coming back to me with some suggestions about changes on material … It's always my choice, you know, they always say, 'Look it's up to you, we don't have to do this.' They're not at all forceful about it, but they do kind of set up a relationship whereby this will mean it will be less expensive than therefore, we will sell more.

In that example, the material choices are potentially nudged by emphasizing the economic impact of the selection process.

Ruben Hutschemaekers, Head of Marketing and Communication at Magis, observed that the designer's agency over material choice is also constrained by the material and technology available to them:

> The designer has maybe the leading role in it … but the designer is also limited to what the industry is offering. I mean, he's not inventing the material he's not inventing the technology to produce it, maybe he's leading the orchestra, but all the different instruments need to be played by [themselves].

This consideration is particularly applicable to renewable plastics, where the complexity of the material is reflected in the wide variety of polymers available, and selection requires expert advice. Bioplastic production is in its infancy and access to supplies varies greatly by geographic region, again impacting choice. Furthermore, manufacturers' access to or willingness to invest in production equipment can also limit the available material choices.

Many of the designers I have interviewed are well-versed in the benefits of working with plastics and keen to communicate the unique combination of attributes that often make them the only suitable candidate for a design project. Lizaso commented:

> We had a meeting last week with another client and they have a mono shell thing, [they want to] produce. But they don't want it to be in plastic … they want for it to be strong; they want for it to be economically viable to sell. Those inputs don't make a result yet, I think.

Plastics are often the only material which can fulfil all the requirements of a design brief at an accessible price point. Barber and Osgerby highlighted the economic advantage of the material:

> We try to use less and less plastics … The problem is wooden chairs are really expensive and are for a high-end market …. We want to design simple everyday chairs, but no one wants to spend a ton of money on a simple everyday chair.

Barber and Osgerby's comment highlight the need to meet a price point, while both quotes illustrate the fact that plastic is often not only the most appropriate solution to a design problem but is also often the only material capable of delivering a project with the constraints of the brief.

i. Specifying renewable plastics

Barber and Osgerby explained the materials section process for their *On & On Stacking Chair* (2019) which is made from a plastic composed of 70 per cent recycled PET from waste plastic bottles reinforced with wood (Figure 5.24). The designers confirmed the manufacturer, Emeco, commissioned them specifically to create a design using that material. In this case the designers benefited as they were absolved from all responsibility to investigate the mechanical properties of the material:

> We work with manufacturers that have a lot expertise and have good partners for materials. Emeco didn't develop the material themselves they worked with a plastic company [BASF] and ultimately, we have to rely on them. [Emeco] tell us what the performance of the plastic is, how it is and we run simulations on the computer and they invest in the moulding and it if does not work it is their responsibility. We delegate the responsibility to Emeco—they are so professional; they research and research. We are very fortunate that we are freed from that side of it as they cover it off.

Their claim was supported in an interview with Buchbinder, who also saw it as his responsibility as the manufacturer to explore new, sustainable solutions for production. Importantly, in this case, the furniture manufacturer always intended to use a recycled material for the project. While the precise material specifications were defined in consultation with an expert material manufacturer, the designer was not directly involved.

Figure 5.24 *On & On Stacking Chair*, Barber Osgerby for Emeco, 2019. Courtesy Emeco.

There are other examples where manufacturers have taken the lead in material specification and commissioned leading industrial designers to showcase new developments in plastics, as discussed in the introduction. Additionally, Grcic worked with Magis to develop the *Bell Chair,* commissioned to utilize recycled plastic, in this case primarily sourced from the manufacturers own waste stream (Figure 4.7). Kartell and Autodesk commissioned Starck to develop a chair using artificial intelligence (AI), aiming to minimize the quantity of recycled plastic needed to match the strength requirements of a chair (Figure 5.21). Hence, both the material and design technique were new to both designer and manufacturer. Starck had previously worked closely with Kartell to successfully introduce PC to the seating market at the end of the twentieth century, again in direct response to the manufacturer's material specifications. These examples demonstrate the manufacturers' drive to innovate and promote the use of new materials.

In contrast, Rashid provided an example of how product designers can influence the choice of renewable plastics when developing a chair. Explaining that he approached the company A Lot of Brasil to develop the *Siamese Chair* (2014) using a plastic injection of the Amazonian fruit Acai and (renewable) bark from the Ipe Roxo tree, probably the first commercially available bioplastic chair (Figure 0.3). In an interview for this book, Rashid stated that he wanted to work with this specific bioplastic and waited for a manufacturer to identify a suitable supply before proceeding with the project. Although Rashid noted that in many cases the manufacturer will be more knowledgeable about the material choice than the designer, and usually takes the leading role in material selection. Only in about 10 per cent of cases, according to Rashid, can the designer have the final say.

Research participants displayed different attitudes to pioneering the use of renewable plastics. Pot reported that he did not want to be the one at the forefront of using the latest materials or techniques just for the sake of being first. Others saw the need for someone to be first and experiment with these materials, despite the risks, as Lizaso explained:

> There was a big possibility, a high risk for the whole exercise not to be something exemplary. But we felt that this kind of jump was needed so that in some iterations someone can arrive to something that is really cohesive and kind of exemplary.

Along with the risks inherent in experimenting with a new material come significant costs which are discussed in more detail in Chapter 7. Diaz, Business Development Manager at Bambacore, an Australian supplier of bioplastics, shared that in her experience, designers are excited to see and learn about the materials available. Translating that excitement to sales is challenging, as the costs inherent in developing new manufacturing processes are often considered prohibitive.

f. End-of-life

Finally, by considering what will happen to plastics products at the end of their useful life, designers can promote more sustainable outcomes. Design for disassembly principles should be applied to promote the recycling of products at the end of their useful life. In addition to simplifying separation of components and materials, designers can ensure that all plastic components are made from polymers that can be easily recycled. Plastic components need to be permanently labelled to facilitate economically viable sorting and recycling, using the internationally recognized RICs. Several of the designers interviewed for this book highlighted disassembly as an important consideration in their work. For example, Dan Armstrong highlighted that many of his clients are concerned about this aspect of sustainability:

> We definitely design thinking we want to be able to pull out the parts and recycle it. I think that's something clients care about. When we design something for them, they want to know that each part can be pulled apart, people are aware of that.

Operating in a global market adds further complexity to the challenge. Armstrong highlighted that some compounds might be suitable for recycling in some markets while other jurisdictions lack the infrastructure to process the same material. Designers might not have complete visibility over where their projects are destined to be sold. Even when the destination is known, detailed information on the waste and recycling infrastructure available in those markets might not be readily available.

2. Summary

Tony Fry argues that driven by a deterministic economic imperative, design serves an instrumental mode of making that brings things into being without knowing what the consequences will be (Fry, 2009, p. 26). As the environmental emergency accelerates, failing to consider the consequences of design is becoming increasingly unacceptable. Designers working with plastics are also increasingly being called to account for their role in the plastics waste crisis by those working on the front line. Parley for the Oceans is an organization dedicated to raising awareness of the beauty and fragility of oceans. Founder, Cyrill Gutsch pointed the finger of blame squarely at the industry reporting:

> I really see it as the obligation of the creative industry, to really reinvent all toxic materials within the next ten years. We have to put a strong focus on that and say, we don't want to create and we don't want to produce products anymore that contribute to the destruction of our planet.
>
> (Marchese, 2020)

Gutsch's comments put the spotlight on the designer. With the equivalent of a garbage truck load of plastic being dumped in the ocean every minute (Ellen MacArthur Foundation, 2017, p. 12), public frustration with the continued expansion of plastic production can be expected to increase. All actors across all industries using virgin plastics will undoubtedly face growing criticism from Gutsch, Fry and many others.

While the ultimate decision on the specific materials to be used for a project might be taken by other actors, the knowledgeable designer can advocate for renewable plastics. Across the product-development process designers are closely involved with a range of decision which can significantly impact the environmental performance of their creations. Most significantly, for products using plastics that do not consume resources during use, minimizing the amount of material will deliver the biggest environmental benefit. Dematerialization efforts also pay dividends throughout the product's life cycle, through reduced demands on the transport and waste industries. Transport efficiencies can be further enhanced by optimizing the final form and associated packaging to minimize resources consumed through distribution.

Many renewable plastics have different characteristics compared with virgin plastics: they often look and feel different and process unique physical properties. Aesthetic disparities can significantly influence a product's appeal. It is the responsibility of design to embrace and celebrate the unique sensory experiences that can be created using renewable plastics. Delivering products which successfully appeal to end-users and sell is equally, if not more, important than meeting conventional benchmarks of sustainable design. In many cases, working with renewable plastics opens new horizons for designers prepared to experiment with the novel forms, colours and textures offered by these materials.

Individual designers and manufacturers are already experimenting with renewable plastics and many have successfully improved the environmental impacts of their creations. This proactive engagement signifies a pivotal step towards sustainable manufacturing. However, the impending challenge lies in the systematic evaluation of the efficacy of these product-level enhancements and their potential for substantial influence on the prevailing demand for conventional plastics. The next challenge is to identify strategies for scaling-up successful initiatives, thereby promoting their widespread adoption and contributing to a meaningful reduction in the demand for traditional plastics.

6 The time for change is now

Those involved with creating products including chairs using renewable carbon-based plastics (renewable plastics) should be commended for their bravery in experimenting with alternatives to virgin plastics. However, most plastic products are still made from virgin fossil resources. While design solutions are often revised to minimize the use of resources, the financial and environmental benefits delivered by this focus on dematerialization are diminishing. Incremental improvements can deliver savings in both material and energy consumption but fail to address the underlying challenges of developing a sustainable consumption model. Far from being confined to the furniture industry alone it has become increasingly apparent that efforts toward sustainability through product level efficiency gains are unable to compensate for the growth in demand from an increasingly populous, affluent and urbanized population (Ryan, 2009). As design academic Ezio Manzini observed:

> Current products and services, taken one by one, use far less energy and materials than those of some decades ago. However, no indicator of aggregate consumption (residence, mobility, tourism, etc.) indicates a decrease: even in countries where research on eco-efficiency has been most successful. Overall consumption of environmental resources continues to increase. This clearly tells us that increasing improvements in the current system are not enough.
>
> (Manzini, 2009, p. 8)

Driven primarily by economic considerations and a desire to maximize profits, a focus on cleaner production and efforts to reduce the total life-cycle impact of products has failed to address ever-increasing levels of consumption and the mounting environmental damage caused by the prevailing linear consumption model. Efforts to reduce resource consumption through eco-efficiency are not enough, as overall resource consumption continues to rise, even, as Manzini highlights, in countries where eco-efficiency has been most successful.

While a system addicted to profit and growth has driven a focus on efficiency it has prevented, or at least hindered, the adoption of designs with long-term ecological benefits that do not provide immediate financial returns (Boehnert, 2018, p. 19). Innovation at the product level is insufficient to drive the change required to our established linear systems of production and consumption. The focus must shift from artefacts and broaden to consider the socio-technical systems requiring revisions in the overall configuration of transport, energy and agri-food systems. However, a single actor, a designer or even a manufacturer of an artefact face a daunting task if they seek to nudge a socio-technical regime toward a more sustainable and resilient orientation, let alone the extractivist regime on which all others depend, energy.

This chapter introduces Multi-Level Perspective (MLP), one of several theories developed to analyse the transition to sustainability through a holistic socio-technical approach. MLP theory posits that new technologies have a greater chance of success when the incumbent industry is in a state of disruption, as entrenched participants become more receptive to new solutions. The exogenous events disrupting the market for virgin plastics are then examined to assess if the time is right for change in the petrochemical industry.

1. Multi-level perspective

The realization that we must embark on a great transition whereby humanity adapts to living within the limits of the planet is gaining momentum. Only one hundred papers in peer-reviewed journals on sustainable transitions were added to the Scopus database in 2000 compared with five hundred in 2018 (Köhler et al., 2019, p. 3) and over 4,700 in 2023. This growing research interest is driven by the recognition that many environmental problems, including climate change, represent grand societal challenges that require system-level transformations to achieve an eco-friendly, circular and carbon neutral economy. There is, for example, little environmental benefit in developing or purchasing recyclable products if the systems and markets are not in place to recover, process and reuse recycled materials.

Systems need transforming so they are responsive to environmental signals and ecological principles (Elzen, 2004, p. 50). The theories developed to address these issues involve a network of actors that extends well beyond the manufacturers and designers required to implement product-level theories, thereby introducing the socially and culturally complex need for networking sustainable innovations (Escobar, 2018, p. 123).

Arturo Escobar observed that transition discourses share the 'contention that we need to step outside existing institutional and epistemic boundaries if we truly want to strive for worlds and practices capable of bringing about the significant transformations [required]' (Escobar, 2018, p. 39). A radical cultural and institutional transformation is required to transition to an altogether different world. Similarly, Tony Fry and Adam Nocek argue that transition cannot be achieved if design limits 'itself to changing the patterns of human consumption, to altering supply chains, or to sourcing biodegradable material' (Fry & Nocek, 2021, p. 3). Transition studies should 'focus on transforming entire technology regimes, rather than separately analysing and prompting specific artefacts or practices' (Elzen et al., 2004, p. 50). The important role that design, in its broadest sense, can play in transitions is increasing emphasized by these writers.

The links between theories describing design interventions aimed at the product level often do not align or have clearly defined relationships with those pertaining to the socio-technical level. MLP is one theory attempting to address this discrepancy by advancing our comprehension of how sustainable innovations emerge, potentially leading to the development, transformation or reconfiguration of existing systems. The theory was originally postulated by Arie Rip and René Kemp in 1988 and refined by Frank Geels and Johan Schot as a model to understand the diffusion of technology (R. Kemp & Rip, 1988; Schot & Geels, 2008). Most transitions studies investigate a single case study rather than attempting to reveal more generic insights (F. W. Geels & Turnheim, 2022, p. 46). My interest is in evaluating MLP as a predictive tool that can assist those interested in driving the ongoing transition to renewable plastics, by identifying and encouraging conditions that are conducive to their acceptance.[1]

MLP is an interdisciplinary model that recognizes that technological transitions (major, long-term technological changes in the way societal functions are fulfilled) do not result from technological innovations (a reformist approach) alone but require changes in, 'user practices, regulations, industrial networks, infrastructure, and symbolic meaning or culture' (F. W. Geels, 2002, p. 1257). Such changes are usually slow, unfolding over decades. Significantly, MLP provides a bridge between evolutionary economics and technology studies. It acknowledges that society-level changes, while desirable, are beyond the influence of individuals or individual organizations. However, MLP operates above the level of individual technocentric improvements by seeking to scale-up and scale-out successful experiments with environmentally superior solutions, thereby influencing technological regime change. MLP focuses on the role of socio-technical systems in achieving societal transformations, differentiating the theory from the debate at the macro level (changing the nature of capitalism) or at the micro level (changing individual choices and motivations) (Köhler et al., 2019, p. 2).

MLP consists of a nested hierarchy of three analytical levels: landscape, regime and niches (Figure 6.1). The landscape level refers to the external socio-technical environment that is beyond the influence of individual actors in the short term (F. Geels, 2007, pp. 402–405). Globalization, climate change, population growth, international trade rules all form part of the landscape. Stable landscape conditions support the ongoing status quo enjoyed by the entrenched social-technical regime.

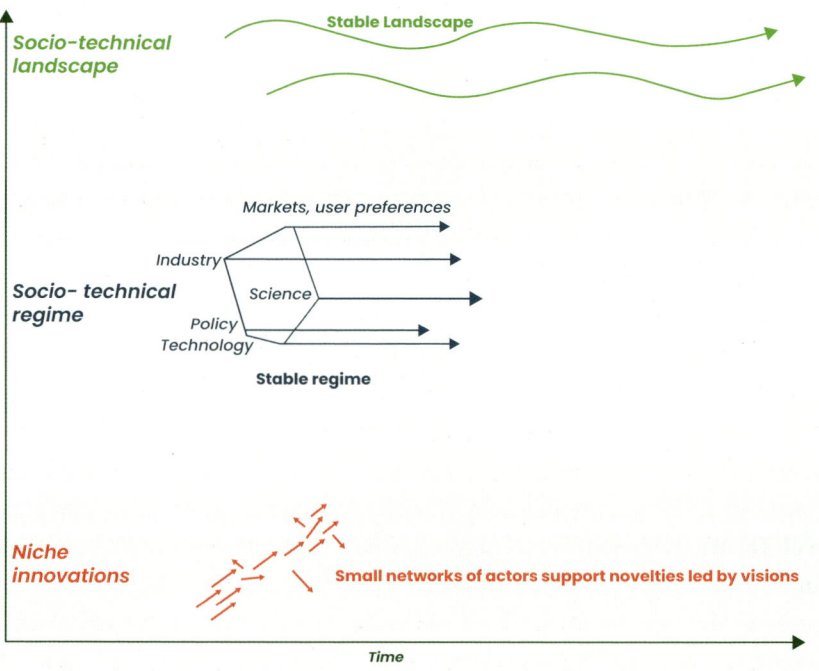

Figure 6.1 The three concepts forming the basis of multi-level perspective theory. Adapted from Schot & Geels (2008).

The socio-technical regime refers to the established rules and entrenched infrastructure that shape engineering practices, production processes, product characteristics, skills, personnel and actor-networks. It encompasses the conventional methods of handling relevant artefacts and defining problems. These elements have developed over decades and are deeply embedded through institutions and ongoing investment (R. Kemp & Rip, 1988, p. 340). Stable regimes do generate innovations but they are usually incremental rather than radical; marginal improvements gained by reconfiguring resources within the confines of the existing regime (El Bilali, 2019, p. 74). Radical innovations are generated in niches, the third concept in the model. Step-change innovations are incubated in environments with sufficient resources to encourage experimentation with technological novelties (Bui et al., 2016). Often emerging from outside or from the fringes of an existing regime, niches foster innovations that differ fundamentally from the prevailing regime. Niches can produce tangible demonstrations that a new technology works to some degree and pique the interest of other industry participants (Smith, 2007, p. 428).

MLP attempts to theorize how such desirable niche innovations can be nurtured to gain the necessary momentum to successfully breakthrough, challenging the status quo enjoyed by the entrenched regime. The theory recognized that transitions cannot be managed in the strictest sense, as they cannot be steered by a single actor. Transitions emerge through a co-evolutionary process involving negotiations between many actors, industry, regulators and activists among them. The theory relies heavily on niche technologies emerging through a process of diffusion based on learning and networking, enabling actors to build on the successes of other industry participants and learn from their failures.

Early critics of MLP noted the complexity and subjectivity involved in defining the socio-technical regime and its boundaries (Berkhout et al., 2004, p. 54; Genus & Coles, 2008, p. 1440). The electricity regime could be examined at the primary level (based on fuel source) or across the entire system of production, distribution and consumption. 'What looks like a regime shift at one level may be viewed merely as an incremental change in inputs for a wider regime at another level' (F. W. Geels, 2011, p. 31). Similarly, while recycle plastics and bioplastics can be considered the innovations, widespread adoption is dependent on their suitability for a wide range of applications. MLP, then, requires the empirical levels of analysis to be designed before the theory can be operationalized.

Monobloc plastic chairs are a niche within the plastic chair market, itself a niche within the plastic furniture market, estimated to be worth between $15 billion and

$23 billion (in 2018), which, in turn, is a small niche within the furniture regime which was worth $677 in 2022 (Grand View Research, 2022b). Furniture is a small niche within the consumer goods regime. As discussed in Chapter 3, production of consumer goods accounts for 11 per cent of the $570 billion worth of plastic produced by the petrochemical regime (Fortune Business Insights, 2023). Plastics, then, represent an important niche for the petrochemical regime, which generated $671 billion in 2022 (The Business Research Company, 2023c). Once chairs (and any other products) are produced their distribution becomes a small niche for the $7.6 trillion transport regime (The Business Research Company, 2023a). At the end of its useful life, a plastic chair, and any product or waste biproduct, if disposed of responsibly, will require the services of the $661 billion waste regime (The Business Research Company, 2023d).

Production of virgin fossil plastics is dependent on inputs from the extraction, mining and energy regimes. The $4.1 trillion extractive regime supplies oil and gas to the energy regime and also supplies the raw ingredients needed to produce plastic to the petrochemical regime (The Business Research Company, 2023b). The $2.1 trillion mining industry supplies coal, used as a feedstock by some plastic producers (particularly in China) and all the resources needed to develop the infrastructure used to manufacture materials and end products. All regimes are niche markets for the energy regime, which generates $2.8 billion profit a day, or a trillion dollars a year, partly by generating energy, mainly by burning fossil fuels (Verbruggen, 2022a, p. 4). This complex network of regime interdependency is illustrated in Figure 6.2.

An analysis of any product is likely to produce a similarly result, often made more complex by the addition of materials sourced from the mining and forestry regimes. The consumer goods regime can be conceptualized as an infinite cascade of niche markets. Importantly, this is not conceptualized as a multiverse, each operating in a separate dimension, but overlapping and intersecting within the same universe, creating the opportunity for networks to develop to share knowledge and learn. MLP was developed through post-analysis to examine how technological innovations succeeded in transforming industries at the high level (for example, Berkhout et al., 2004; F. W. Geels, 2002). I examine if the theory can explain transitions at a micro level. The objective is to ascertain whether the theory remains effective in explaining transitions within such a specific niche. Can a comprehensive analysis of the drivers and obstacles encountered by innovators in one sector provide guidance to stakeholders engaged in production of consumer goods more generally? By extrapolating from this micro-level study, a broader and

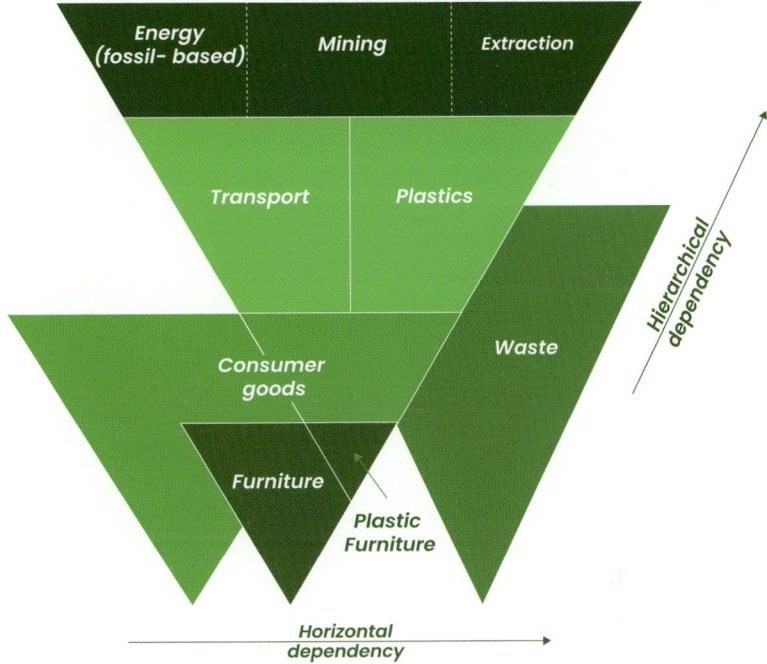

Figure 6.2 Regime dependency for fossil monobloc plastic chairs. Adapted from Gaziulusoy (2010).

more profound impact on the entire consumer goods sector becomes conceivable, potentially making a significant impact on the demand for virgin plastic.

Innovations in the materials market can only become mainstream if they succeed in replacing the use of virgin plastics across multiple applications. Demand for renewable plastics to create packaging is increasing, but packaging accounts for only about one-third (36 per cent) of total demand for plastics (Chapter 3). Plastics are used across many other sectors including construction, textiles, transportation and consumer goods. Renewable plastics need to be adopted as a replacement for virgin plastics across those other market segments to drive the desired change across the entire entrenched regime. An incremental decrease in the quantity of virgin plastics used to make consumer goods will add to the reductions taking place in the packaging market and combine with the efforts of others across multiple sectors. Each sector can be conceptualized as its own regime with its own set of dependencies and network of actors (some unique) and encountering its own combination of enablers and barriers which must be overcome before the new materials are accepted.

The enormity of the challenge to be faced in transitioning to renewable plastics is highlighted by considering the impact on the dependent regimes (Figure 6.3). In this more sustainable scenario, energy is primarily generated from renewable sources. Agriculture (including aquaculture) largely replaces the extractive industries to become a source of raw materials for virgin renewable plastics. The mining regime will, of course, remain essential for their contribution to the development of infrastructure (and essential for the supply of elements, such as rare earths, and lithium, needed to enable other regimes to make sustainable transitions). Meanwhile, the petrochemical regime will be threatened by the waste regime. No longer restricted to offering an end-of-pipe solution, disposing of unwanted items, the waste regime will morph into the resource recovery regime and become an important supplier of fuels and feedstock to other regimes and the primary supplier of recycled plastics. The transformation of the waste industry is crucial in the development of a circular economy (United Nations Environment Programme, 2024).

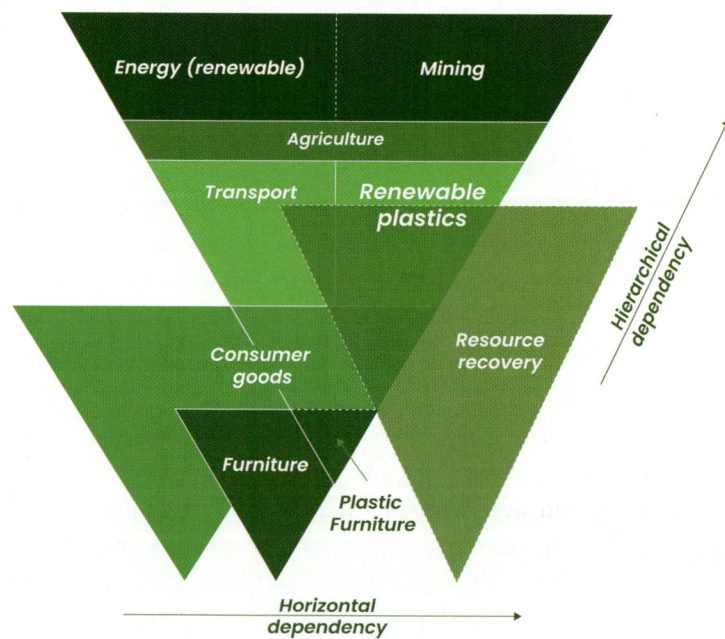

Figure 6.3 Regime dependencies for renewable carbon-based monobloc plastic chairs. Adapted from Gaziulusoy (2010).

However, MLP posits that it is not mandatory for an emerging technology to replace the established social-technical regime, incumbent actors can choose to adopt or adapt and reconfigure to accommodate changes. Indeed, in the case of renewable plastics reconfiguration is essential, as the petrochemical regime alone has the infrastructure, financial resources and expertise needed to affect a smooth transition within a realistic timeframe. Manufacturing renewable plastics at scale requires access to refineries and crackers, which are enormously expensive to build (Chapter 3). Successful reconfiguration of the petrochemical regime will depend on the success of new chemical recycling technologies, many of which are themselves in the early stages of niche experimentation (Chapter 7). These activities can be viewed, through the lens of MLP, as the petrochemical regime experimenting with their own niche solutions with the aim of reconfiguring to retain their dominance over the plastics regime. This shift in focus from a bottom-up disruption caused by niche innovations replacing the existing regime to a reconfiguration process is consistent with the findings of a study of transitions occurring in the electricity, heat and mobility regimes in the UK (F. W. Geels & Turnheim, 2022, p. 293).

Some might argue that the fastest way to affect the desired change in plastic production change it to accelerate the manufacture of renewable plastics at scale. However, simply creating and supplying more renewable materials does not automatically create demand for them. Additionally, MLP studies have shown that challenger technologies need to be sufficiently mature if they are to succeed in replacing the incumbent regime. Henry Ford experimented with soybeans and produced a prototype for a car made from bioplastics in 1941 (The Henry Ford Museum, n.d.). Following the end of the Second World War, an abundance of inexpensive petroleum spurred investment in the rapidly developing petrochemical industry and Ford's plastic car never went into production.[2] In response to the oil crisis of 1973, ICI began to develop Biopol, a (PHB) polymer made from fermented glucose. As the oil market faced further disruption toward the end of the decade, with the price of oil remaining high, opportunities for the commercialization of bioplastics appeared ideal. In 1974, Hayaskihara Biochemical Research Institute announced it had 'applied for patents in twenty-five countries for a plastic [starch based] material which dissolves in water, emits no poisonous gas if burned and can be eaten' ('Plastics Made from Starch', 1974). However, once the price of oil dropped in the 1980s interest in developing these alternative plastics quickly waned (Bennett, 2012a, p. 332). Industry chose to focus on pursuing a programme of efficiencies, dematerialization reigned, while the opportunity to replace fossil fuels as a feedstock was ignored (R. P. M. Kemp et al., 2001, p. 279). The low price of oil effectively de-incentivized development of alternatives to fossil fuels and 'insights were too limited to prevent neoliberalism' (Verbruggen, 2022b). By focusing on short-term profits and ignoring mounting evidence highlighting the external environmental costs caused by the continued use of fossil fuels, the prevailing fossil regime withstood these disruption and regained stability while continuing to produce ever-increasing volumes of plastic.

To drive the uptake of renewable plastics in consumer goods, it is important to not only demonstrate their technical suitability but also gain acceptance by consumers and business purchase decision makers. In Chapters 4 and 5, recent experiments with renewable plastics to develop seating solutions were examined; similar innovations are occurring across many other categories of consumer goods. These niche innovations are represented at the bottom of the chart (Figure 6.4). With no recognized superior solution yet identified,

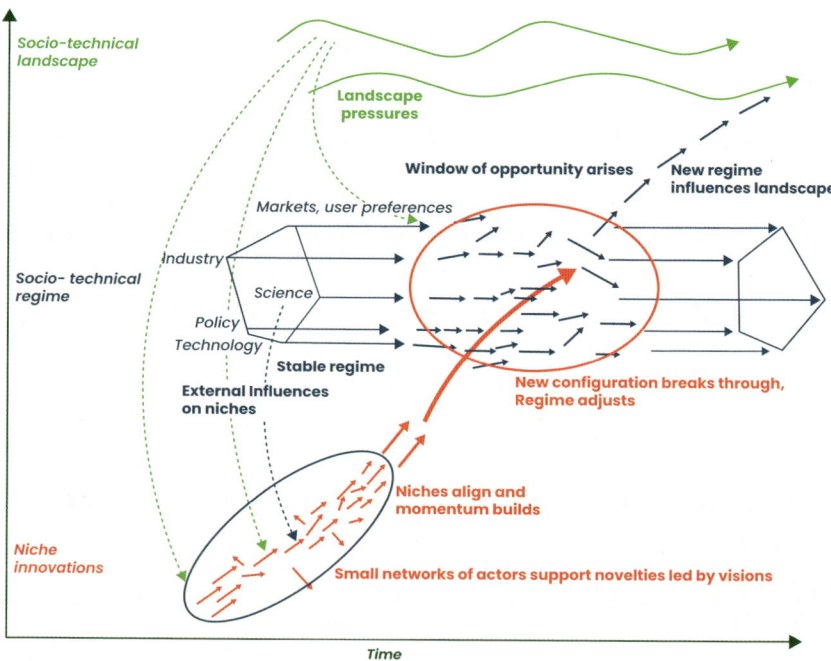

Figure 6.4 MLP model illustrating how niches can take advantage of landscape shifts or resultant tensions within the regime to breakthrough. Adapted from Schot & Geels (2008).

various alternatives are being experimented with (represented by arrows pointing in random directions at the niche level, circled in blue). Some of these niche experiments will fail.

Ezio Manzini argues that a pathway to transition can be formed by experimenting with new solutions, consolidating and replicating the best ones. Sustainable design evolves through sharing of knowledge about successful interventions, which could result in the development of new market niches. Good design solutions are imitated, successful strategies copied, and these new projects can create more effective solutions. Eventually these solutions coalesce, enabling the multiplicity of small initiatives to make a greater impact (Manzini, 2015, p. 5). It is this process of learning from real-life experiences which, Manzini argues, is the main design strategy to change complex systems. Favourable eco-systems can be developed to enable new innovations to flourish, scaling-up and scaling-out as successful innovations gather support and interest from other actors (Manzini, 2015, p. 59). Over time, a dominant or preferred solution might emerge from the niche experiments through trial and error, with industry participants learning from both successes and failures.

MLP theory anticipates significant friction resisting change from a variety of factors, including pre-exiting standards and regulations, technical systems, sunk investments in machines and equipment, infrastructures, employee skills sets and competencies (examined in the next chapter). All these factors form the entrenched socio-technical regime, with participating actors in a balanced symbiotic relationship, with well-established routines and relationships (Davidson et al., 2016, p. 368). Nevertheless, all socio-technical regimes have internal contradictions or tension reflecting the varied interests of stakeholders that causes destabilization and creates opportunities for change to occur (Fuenfschilling & Truffer, 2014, p. 772). In Figure 6.4 the resulting weakened links between these participants of the regime are represented by short diverging arrows at the regime level (circled in red). These shifts or tensions can also create opportunities for niche innovations, as they pressure the regime to search for alternative solutions.

Similarly, dramatic disruptions at the landscape level can create opportunities for niche innovations to breakthrough by exasperating tensions and creating disturbances within the regime, again prompting a search for new solutions. It is then important to examine and understand the exogenous landscape events causing major disruptions currently impacting the petrochemical industry. This is particularly important as every application of renewable plastics is impacted by landscape conditions.

2. Landscape disruptions

At least five major components of the landscape level are experiencing disruptions, creating tensions within the existing fossil plastic social-technical regime: China Sword, climate change, health concerns, government intervention and price shocks (Figure 6.5). While all five of these landscape disturbances are exerting pressure on the existing regime there can be no guarantee that the disruption is sufficient for a challenger technology to breakthrough and achieve widespread

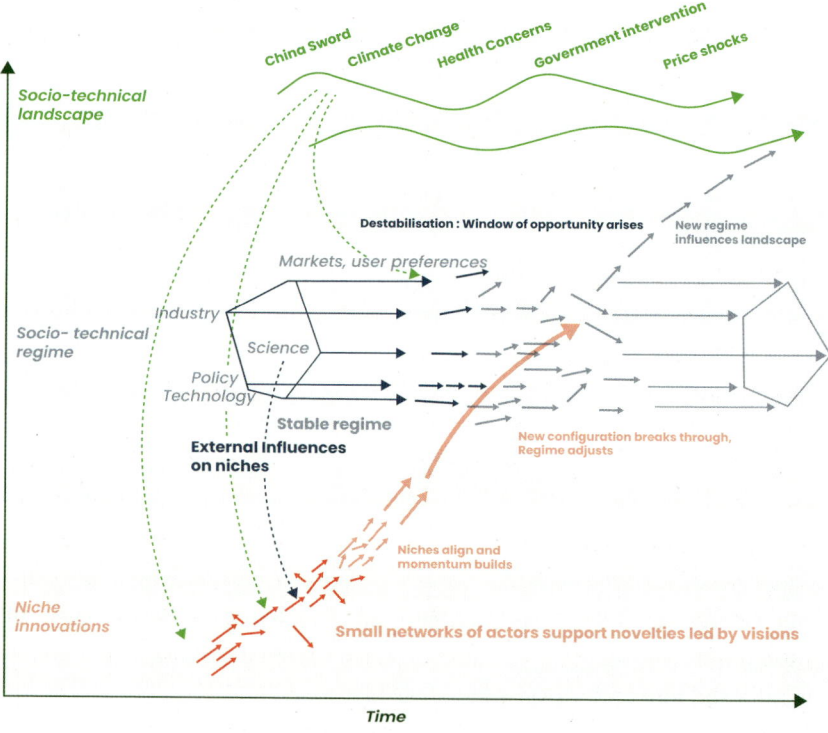

Figure 6.5 Landscape pressures emerge and create tensions with the existing socio-technical regime. Adapted from Schot & Geels (2008).

adoption or that a niche can gather sufficient momentum to take advantage of the opportunity (see for example the previous attempts at breakthrough by bioplastics discussed above). But major disruptions at the landscape level can create opportunities for niches to gain momentum as industry participants seek to mitigate their impacts (Smith, 2003, p. 131).

At the start of 2018 China banned the import of mixed waste plastics. Just as the enormous impact of China's ban was beginning to be realized the issue was dwarfed as the world confronted a pandemic to be followed by war in Europe, adding to price disruptions. Meanwhile, mounting evidence of both climate change and plastic pollution are being met with increasingly urgent demands for action from politicians and activists alike. An analysis of the landscape today reveals the magnitude of disruption currently confronting the petrochemical regime.

a. China Sword

China caused the first major disruption to the plastic market since the GFC, with the introduction of the National Sword (China Sword) policy at the start of 2018, effectively banning the import of mixed plastic waste. For decades high-income countries dealt with their plastic waste problem by exporting it to the global south (waste colonialism). In response to China's new policy, some countries (including Australia) redirected their waste to other Asian nations including Indonesia, Thailand, Malaysia and Vietnam. This response proved to be a short-term solution, as these countries swiftly introduced similar bans forcing Europe, Canada, the United States, Japan and Australia to finally confront their own plastic waste issues (Coca, 2020).[3] The Australian Federal Government has sought to circumvent international regulations to prevent the export of mixed plastic waste simply by reclassifying the waste as Refuse Derived Fuel (RDF) or Processed Engineered Fuel (PEF), allowing the country to side-step its obligations (Reuters, 2022). This has led to the allegations that 'Australia is acting against the spirit of the Basel Convention by claiming that RDF is not waste' (Bosch, 2022) and 'Australian waste export ban announcement amounts to little more than a public relations exercise to maintain waste movement out of Australia and into less wealthy countries' (Bremmer, 2022, p. 8). While Australia did export less than half the amount of plastic waste during 2023 compared with the years leading up to the implementation of China Sword (64.9 million kg compared with 150 million kg), the majority (61.1million kg) was still sent to non-OECD countries (Basel Action Network, 2024).

It will take years for affected countries to develop strategy, arrange finance, negotiate feedstock arrangements, obtain planning permissions and build the infrastructure required to significantly impact the proportion of plastics recycled.[4] The CSIRO estimate Australia needs to increase recycling capacity by 150 per cent to compensate for the lost export markets, which only ever processed up to 12 per cent of the plastic consumed within the country (CSIRO, 2021a, p. 14). The EU is doing slightly better, having increased its recycling capacity by 17 per cent to 11.3 million tonnes in 2021, sufficient to process about 18 per cent of estimated annual consumption (Amadei et al., 2022; Laird, 2023). However, to reach its recycling targets, the EU requires investment of between €6.7 and €8.6 billion in sorting facilities by 2025 and a further 4.6–5.7 billion to increase recycling capacity (European Investment Bank, 2023, pp. 23–25). Meanwhile stockpiling, landfilling and waste to energy (all less desirable than recycling) are becoming increasingly popular as short-term strategies to combat the increase in domestic waste.

On the positive side, China's stand against waste colonialism acted as a catalyst, focusing attention on the negative environmental impacts of plastics, particularly packaging and SUPs. The China Sword policy has resulted in the established regime being disrupted and challenged and might yet prove to be the single most significant landscape avalanche event, driving a re-evaluation of both the production and our consumption of virgin plastics.

b. Climate change

Unlike conventional economic analysis the MLP model recognizes that broad political and social attitudes and trends play a crucial role in shaping and changing socio-technical regimes (Smith et al., 2005, p. 1494). Scientific consensus has emerged over the link between anthropogenic GHG emissions and climatic change. Increasingly extreme weather events are demanding a reinterpretation of the reality created by the existing regimes and further fuelling demands for change (Fuenfschilling & Truffer, 2014, p. 776). The shifting cultural discourse is delegitimizing the fossil fuel industry.

Growing political pressure for action on climate change, epitomized by the increasingly militant activities of groups such as Extinction Revolution and Just Stop Oil, add to the mounting scientific evidence pressuring international regulators to develop decarbonization policies, reducing reliance on carbon-intensive energy sources and promoting energy efficiency. As the supply of renewable energy increases and demand for fossil fuels slows, further pressure will be exerted on petrochemical companies to maintain their profitability. Demand for oil is forecast to remain flat or increase only slightly during the next decade (Center for International Environmental Law, 2019, p. 12). To help compensate for these losses the plastics market is seen as a critical growth segment for the petrochemical industry. Investments are being made to expand production to satisfy the projected 3.5 per cent CAGR through to 2050 and beyond. Reduced demand for fossil fuels has paradoxically amplified the importance of plastics to the extractive and petrochemical regimes, who now regard these materials as critical to sustaining profitability.

Current industry projections forecast demand for fossil plastics to triple by 2050, which is widely regarded as unsustainable (Ellen MacArthur Foundation, 2016, p. 24). Growing public pressure, international agreements and increasingly restricted access to finance for fossil extraction activities will eventually force even the most resistant governments and companies to take more action to combat climate change. The GHG benefits offered by renewable plastics will likely be increasingly recognized as an important tactic in our response to the climate emergency.

i. Environmentalism

Discarded plastics end up in our environment killing wildlife and degrading soil fertility. Growing concerns, particularly about ocean plastics, have fuelled public discussion and focused demands for action, expressed through groups such as Plastic Free July (Chapter 3). These same concerns have led to a wider debate about the role of plastics and demands for change at the socio-technical level. This increase in concern for the ecological impacts of plastic represents a significant cultural shift.

Environmental concerns are not only expressed through political protests and petitions; environmental lobbying can be a powerful tool to undermine support for incumbent regimes (Fuenfschilling & Truffer, 2014, p. 781). Public opinion is often monitored through market research, with results considered by organizations to guide strategy. A 2019 survey conducted across twenty-eight countries (IPSOS, 2019) found public support for governments to ban SUPs and introduce extended producer responsibility (EPR) schemes (Figure 6.6 shows detailed results). Over 70 per cent of participants agreed they felt better about brands that made changes to improve environmental outcomes and wanted to support efforts to minimize packaging.

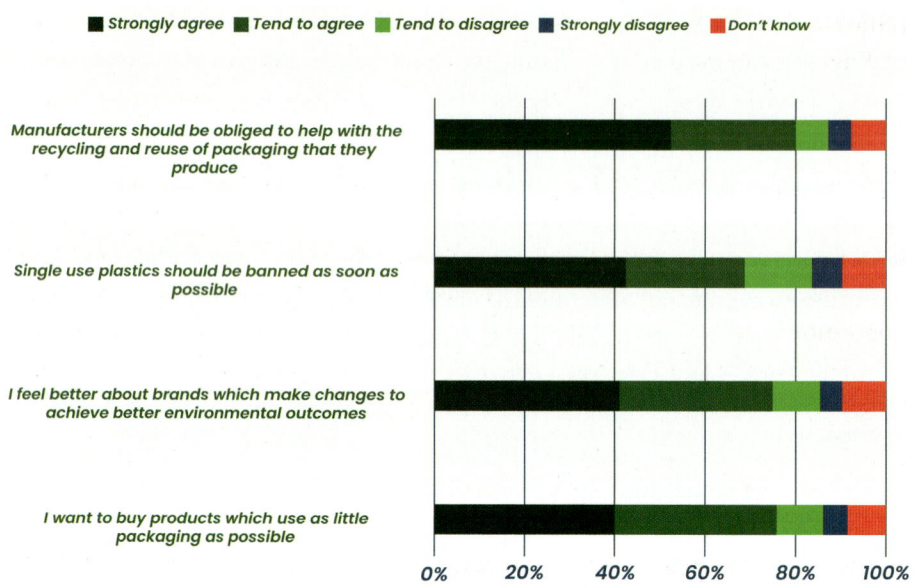

Question : Do you agree with the following statements?

■ Strongly agree ■ Tend to agree ■ Tend to disagree ■ Strongly disagree ■ Don't know

Manufacturers should be obliged to help with the recycling and reuse of packaging that they produce

Single use plastics should be banned as soon as possible

I feel better about brands which make changes to achieve better environmental outcomes

I want to buy products which use as little packaging as possible

0%   20%   40%   60%   80%   100%

Source: IPSOS (19,515 online adults under the age of 75 surveyed across 28 countries. Fieldwork dates: July 26th – Aug. 9th, 2019)

Figure 6.6 International survey of attitudes toward packaging (2019). Adapted from IPSOS (2019).

Tony Fry emphasizes the importance of making visible the unmaking caused by products such as straws as key to changing consumer behaviour. Once people became aware of the devastating impacts straws have on our environment and on wildlife, they are motivated to alter their consumption patterns (Fry, 2020, p. 78). Organizations such as the Ellen MacArthur Foundation are raising public awareness of the unmaking caused by plastics, particularly SUPs and advocating for producers of FMCGs to transition to more sustainable alternatives.

Focusing attention upstream on the devastating unmaking caused during the manufacture of plastic is also driving change. Mounting evidence demonstrates that environmental concerns are causing tension, weakening faith in the incumbent fossil regime. Increased community environmental concern and action around plastics has already directly impacted the development of at least one planned fossil plastics plant in the United States. The Louisiana Bucket Brigade is an action group dedicated to partnering 'communities to help residents amplify their voices and challenge the petrochemical industry's relentless expansion' (Louisiana Bucket Brigade, 2020). In November 2020, this 'not-in-my-backyard (NIMBY)' resident action group succeeded in persuading the US Army Corps of Engineers to suspend the permit for the Formosa Plastics Plant in St James Parish, Louisiana, planned to be the largest petrochemical plant in the United States. In fact, the proposed facility expected to emit 800 tons of GHGs each year, doubling the current level of emissions from the entire St James Parish. The group opposed the project as 'the plant would pollute a predominantly Black community, disturb unmarked burial sites of enslaved people, degrade wetlands and add to the ocean plastic pollution crisis' (Louisiana Bucket Brigade, 2020). Although opposition was motivated by a range of community concerns, the petrochemicals expansion plans were interrupted, at least in part, due to increased concerns around the environmental damage caused by plastic. In 2023, the Louisiana's 19th Judicial District Court reversed its decision to issue air permits that would allow the Formosa Plastics complex to be built (Trimble, 2022). In January 2024, this decision was overturned on appeal and the fight continues.

The Louisiana Bucket Brigade is not the only local environment group causing disruption to the fossil fuel regime and these efforts were turbocharged in 2022, with an $85 million campaign financed by Michael Bloomberg specifically to resist development of new plastic manufacturing facilities. Increasing community environmental concerns are credited with discouraging international investors from joining the ethane fracking boom in America. Finance professor Kathy Hipple reported: 'more and more we're seeing from earnings calls and financial reports that local opposition has become a material risk factor' (Marusic, 2021). For foreign companies seeking to invest in petrochemicals, understanding the intricate patchwork of local state and federal regulations is complex enough, without the additional threat of lawsuits from community groups.

These examples demonstrate the complex web of actors that are essential to driving a transition. Apart from direct effects these interest groups can also impact the opinions and action of others including product designers. Industrial designer Bertjan Pot revealed to me during our interview that he would no longer use carbon fibre, a material he had previously used in his designs:

> I know for myself that I wouldn't make the *Carbon Chair* again. I mean, I like the chair and I think it's a good thing and they should keep making it because it's a good thing. Worst thing about the product is that you throw it out, but I don't think the *Carbon Chair* is something you throw [away]

easily. So, on that point, it's quite safe, but to recycle, it's horrible, as all composites are when you glue materials together, you know you are not making it really usable in a cradle-to-cradle way.

Speaking more generally about plastics he revealed his growing disquiet with working with the material and its end-of-life prospects:

> For Moooi I did propose another light to them last year, actually a few, when one of them had quite a lot of plastic in it. And then Marcel [Wanders] said, it's too much plastic. I think he's right … Where is this block of light, a block of plastic, ending up through the lifecycle?

Gabriel Chiave, from Marcel Wanders office, confirmed that the studio is actively trying to reduce its involvement with projects that involve plastic:

> If it's not extremely innovative, or if it's not extremely revolutionary in a way because the world doesn't need another huge roto-moulded plastic base or seat. So, we try to avoid those type of projects as much as we can.

Other designers reported that environmental concerns are driving changes in consumer preferences. Barber Osgerby even suggested that the public might abandon some of the more harmful plastics:

> Now everyone has realized that bioplastics are the new way forward. If you are going to continue to produce PC, people aren't going to buy that in a couple of years, if not by next year. Companies like Kartell must seriously be considering their business plan by now. Most other companies probably have a more diverse base of products. It's great, thank God it is happening now.

Likewise, Louis Durot predicted the demise of PU:

> Today I use polyurethane that contains a highly toxical [sic] component before production. After production the product, if not burnt, is not offensive/dangerous, but the components to produce it are dangerous. Just like for Bakelite, my artwork, that I am still allowed to produce now, will be forbidden to produce within a few dozen of years.

Pot even went further, suggesting that environmental concerns are directly impacting consumer preferences:

> I think there's also this almost visual need for things being lighter, more transparent less material. So, in a way like the visual trend is keeping up

with the [environmental] trend or with the real issue, or the practical issue, of why we shouldn't use plastic that much.

These designers, prompted by increased awareness of the environmental harm caused by plastics, are actively reconsidering their relationship with the material. Some have even gone so far to commit to stop using certain types of plastic. Even major companies like Kartell, who have built their entire business around fossil plastics, are actively reassessing their dependence on them, as discussed in more depth in Chapter 8.

These shifts in actor-related perceptions demonstrate how 'mood' can accelerate or slow down diffusion and affect breakthrough (F. W. Geels, 2005a, p. 692). There is risk that a single symptom of the plastics problem (ocean plastics) becomes prioritized in political debate and monopolizes scientific attention to that issue (Nielsen et al., 2020). By focusing on a niche downstream impact attention is potentially diverted from the source of the problem. The challenge is for more efforts to be directed upstream at the root cause of the problem. Every designer or manufacturer who decreases their use of virgin plastic resists the petrochemical industry's expansion.

c. Health concerns

We are eating and inhaling plastic every day, despite mounting concerns of potential health impacts. A report, based on an analysis of fitty-two studies, found that humans are ingesting five grams of microplastics every week which is equivalent to eating a credit card (Dalberg, 2019, p. 6). The leaching of chemicals from plastics to wildlife and humans through the air or food chain have all raised public concerns (Zalasiewicz et al., 2016). A team of nearly fifty scientists working on a comprehensive report documenting the health impacts of plastic concluded, that plastic causes disease, and premature mortality at every stage of its life cycle, disproportionately affecting vulnerable, low-income and minority communities. Toxic chemicals added to plastic are known to increase the risk of miscarriage, obesity, cardiovascular disease and cancers (Landrigan et al., 2023). Elephants are not the only species threatened with extinction by the ongoing production of fossil-based plastics.

A 2009 report from the Centers for Disease Control and Prevention in the United States found traces of bisphenol A (BPA), flame retardants, phthalates and poly-brominated diphenyl ethers in the bodies of Americans (Liboiron, 2013, p. 139).

Researchers have linked exposure to these, and other chemicals, to infertility, with sperm counts among Western men more than halving in the past forty years (Hu et al., 2024; Swan & Colino, 2021). Another study found abundant microplastics (carboxylated polystyrene spheres (PS25C)) within the placenta and in all foetal tissues of pregnant rats examined, including liver, kidney, heart, lung and brain. This study confirms that nanoscale plastic particles can transfer to unborn children, potentially impairing foetal development (Cary et al., 2023). Nanometer-sized PS particles have been shown to traverse the blood brain barrier in mice only two hours after ingestion (Kopatz et al., 2023). Various microplastics have been found in patients undergoing heart surgery (Yang et al., 2023). BPA, a precursor monomer used in the production of PC, can have alarming impact on 'essential communication functions of neurons in mature vertebrate brains', with the researchers calling for the rapid development of alternative plasticizers (Schirmer et al., 2021, p. 1). Plastics simply refuse to go away. Their material recalcitrance is forcing us to acknowledge the ways in which plastics persist in our environment and in our bodies, long after their original use value is exhausted (Thompson, 2013). As the weight of evidence of health impact increases the pressure for change intensifies.

However, health concerns also threaten the expansion of the recycling industry and threaten to hinder the transition to renewable plastics. As demand for recyclates grows, more recycling infrastructure is needed. Proposed new recycling facilities often face NIMBY opposition from local groups justifiably concerned about odours, pollution, traffic, fire risk and potential health impacts. Issues of racial and social justice are also often likely to complicate development applications, as facilities are frequently located in low-income neighbourhoods with high concentrations of people of colour (Heffernan, 2022). While understandable, the success of these resistance activities could well hamper sufficient volumes of recycled plastics from becoming available.

d. Government intervention

Sustainability is a public good, meaning that individual private actors have limited incentives to address it. Government policy needs to play a central role in shaping the directionality of transitions through environmental regulations, standards, taxes and subsidies (Köhler et al., 2019, p. 3). Government can adopt the role of a transition manager, directly affecting the success of a technology or the speed with which it is adopted (Köhler et al., 2019, p. 3). Intervention is particularly useful to support currently financially unviable innovations (such as most bioplastics).

Regime change can be accelerated by protecting and encouraging desirable niche innovations or mandating their use. Macro-level policy guides individual choices at the micro level (Dewberry & Johnson, 2010, p. 2). However, policymakers often effectively form alliances with incumbent organizations making it difficult for fundamental change to occur (F. W. Geels, 2010, p. 502).

A mutually advantageous, symbiotic relationship has developed between governments and the fossil fuel regime. This relationship has been strengthened by an industry-led strategy of lobbying, supported by donations, and a shared alignment of goals between policymakers and businesses due to neoliberal market economics. Policy proposals with a focus on environmental benefits often conflict with sector-specific goals defended by policymakers from other more powerful ministries (F. W. Geels & Turnheim, 2022, p. 39). With the established fossil fuel regime in a powerful alliance with government it is often challenging to implement policy interventions designed to deliver long-term environmental benefits.

Consistent international action is the most efficient method to drive change in the plastics industry. International agreement on definitions, standards and labelling (particularly the bewildering and confusing array of over 450 eco-labels) backed by consistent informational instruments would be an important start but remains elusive.[5] The Basel Convention on the *Control of Transboundary Movements of Hazardous Wastes and Their Disposal* enjoys near-universal support. Mixed plastic waste can only be exported with the prior consent of the importing country (although some countries, including Australia, have already found ways to avoid these restrictions, as discussed earlier).

On the positive side, an attempt to implement a global governance arrangement that addresses the whole life cycle of plastics is underway. The agreement calls for production of virgin plastics to be reduced, with recycled plastic used in products where feasibly possible and for the introduction of EPR schemes. The proposal was presented to the UN Environment Assembly in February 2022, with a first draft of an agreement circulated in September 2023, but it appears unlikely that agreement will be reached in time to meet the original 2024 deadline, with many participants resisting calls to limit production of plastics.

The UN has also involved itself with specific cases, particularly where there is evidence of environmental racism. In March 2021, the UN Human Rights Commission raised serious concerns about further industrialization of 'Cancer Alley', a region in Louisiana that is home to multiple petrochemical complexes.[6] The UN found the proposed Formosa Plastics Plant could increase the cancer risks in predominantly African American Districts in St James Parish to about

105 cases per million, while in other predominantly white districts the cancer risk is in the range sixty to seventy-five per million (UN News, 2021). Environmental justice concerns are reinforced by the fact 90 per cent of GHG emitted from plastic production in the United States occur in just eighteen communities, which are 67 per cent more likely to be inhabited by residents earning 28 per cent less than the average and are 67 per cent more likely to be inhabited by people of colour (Beyond Plastics, 2021, p. 7). But the UN has no power to stop this or any future developments in Louisiana or elsewhere. In this case, local resistance groups have found more success than the UN.

Meanwhile there are international initiatives to tackle the issues of microplastics or toxic chemical additives. The EU has introduced legislation to restrict the intentional addition of microplastics to products such as cosmetics, but this fails to address tyres, and laundry, which are the main sources. Currently, plastic manufacturers face little regulation over their product offerings despite more than 1,500 substances often used in plastics being listed as chemicals of concern (Simon et al., 2021). Fillers, additives and colouring agents mixed with polymers to improve mechanical properties or aesthetic appearance can contain toxins. Despite the health impacts discussed above little international action has been taken to ban the use of these potentially dangerous substances.

At the national level, the politics of plastic have primarily focused on single items with plastic bags, plates or cutlery commonly targeted by legislation in response to public concern. Recently, there is a shift to initiatives aimed at banning all SUPs, the focus of politics is shifting towards the entire system, no longer focusing on single items (Nielsen et al., 2020). Some jurisdictions have already used policy to encourage post-consumer recycling, such as the EU Directive (94/62/EC), requiring all plastics packaging used throughout the EU to be reusable or recyclable in an economically viable way by 2030. Germany has legislated EPR, making producers accountable for waste management, launching the Green Dot scheme to implement recovery and recycling of packaging (Hopewell et al., 2009, p. 2122). The United States proposed a tax of up to twenty cents per pound on virgin plastic resins, but this is being resisted by industry, led by the American Chemistry Council. The UK introduced a plastics tax of £200 per tonne of packaging that does not reach a threshold of 30 per cent recycled plastic in April 2022. The EU introduced a tax of €0.8 per kilograms of non-recycled packaging waste in 2021 ('Europe Is Implementing a Tax on Plastic', 2021). Other jurisdictions are relying on voluntary commitments from FMCGs (such as those being led by the Ellen MacArthur Foundation) to reduce or eliminate their use of virgin plastics.

With the price of recycled plastics now often exceeding that of virgin plastics it is yet to be determined if these voluntary commitments will be honoured in the long term (Paben, 2023).

Despite intense opposition from industry, an increasing number of jurisdictions are introducing container deposit schemes (CDS) to tackle plastic littering. By offering cash incentives to return beverage containers reductions in littering have been reported.[7] Such schemes also facilitate sorting at the point of collection, allowing plastics to be more accurately separated by type and potentially producing cleaner streams for recycling.

Governments can go further and mandate the use of use of recycled plastics and bioplastics in products purchased by government agencies or even by industry. China has led the way here, stepping into the role of an active transition manager, mandating the use of recycled materials by key industrial sectors, including the automobile industry. Government actions could be further supported by information campaigns developed to explain the virtues of renewable plastics to the community.

Following the introduction of the China Sword policy, governments in high-income countries are being forced to intervene to encourage rapid expansion of domestic plastic recycling infrastructure. Grants and loans (and planning shortcuts) are being made available to stimulate development. While cheap finance is popular with the entrenched regimes and is likely to eventually increase the supply of recyclates, it does nothing to stimulate demand. Europe has attempted to address that imbalance through its tax on non-recyclable plastic packaging, thereby creating demand for recycled plastics.[8] This approach mimics petrol taxes (estimated as averaging €80 per tonne of CO_2 globally), which raises revenue by making polluters pay to compensate for the externalities created by its consumption (estimated at over €100 per tonne of CO_2) (Carbon Tracker Initiative, 2020, p. 35). Having used price signals to stimulated demand for recyclates, governments can then distribute the money raised from forcing polluters to pay for externalities such as the grants to build the recycling infrastructure needed to meet increased demand (Carbon Tracker Initiative, 2020, p. 35).

Renewable plastics and bioplastics have not yet received the level of support afforded by governments to assist the biofuels industry. For example, in the 1970s, the Brazilian government committed to purchasing fleets of cars running on biofuels while, in 2003, the EU introduced the Biofuels Directive, both initiatives elements of broader strategies designed to accelerate the uptake of ethanol (Silveira & Johnson, 2016). Support for the bioplastic market has been largely absent, although Brazilian petrochemical organizations continue to benefit from the previous R&D investment in biofuels. R&D could also be supported to develop bio-based chemicals.

Developing a completely new industry takes years or even decades. While many bioplastics show potential in the laboratory, there are no guarantees that production at an industrial scale is technically or economically feasible. Venture capital investors are more likely to support projects with a lower-risk profile and faster, guaranteed payback (Bennett, 2012b, p. 102). The important role government could play in driving demand for bioplastics was illustrated by the industrial designer, Louis Durot, in my interview with him:

> To date no company has asked me to help them to research biodegradable plastics ... because without special laws there is no market or need for fully biodegradable plastic. With time, plastic becoming more and more polluting, probably a legal system will come.

Indeed, China has led the way in introducing policies to promote bioplastics. Fuelled by a desire to break away from its dependency on imported oil, China has ambitious plans to expand production (Greenpeace International, 2020, p. 1). Thailand has also been quick to recognize the potential of bioplastics, offering tax incentives to companies to encourage the use of biodegradable plastic packaging with the aim of persuading 10 per cent of existing plastic manufacturers to switch to bioplastics (Barrett, 2019a).

Importantly, regulations establish goals and provide direction on the environmental outcomes that need to be achieved. Therefore, regulators need to be wary of unwanted consequences of high-level targets or the need for complementary policies to address skills, supply chains, public acceptance or industry support (F. W. Geels & Turnheim, 2022, p. 325). For example, the Australian Packaging Covenant Organisation (APCO) is charged with meeting Federal Government targets, including that an average of 50 per cent recycled content is to be used in all packaging by 2025 (APCO, 2019). In 2018/2019 HDPE containers averaged only 2 per cent recycled content (Chapman, 2020). Industry took the easy route, increasing the use of recycled paper and cardboard, allowing them to meet the overall target, while ignoring the more complex issue of plastics. These actions will make it virtually impossible to reach the government's secondary target of 70 per cent of plastic packaging to be recycled or composted by 2025.[9] Government targets can be better designed to compel immediate action on phasing out virgin plastic packaging.

Government intervention has the potential to quickly create opportunities for designers interested in sustainability. Policy interventions can be designed to go beyond supporting the entrenched waste regime by financing recycling infrastructure in isolation and invest in the development of bioplastics. Demand-generation initiatives, largely neglected to date, should broaden their focus beyond packaging; mandating or incentivizing manufacturers to decrease their use of virgin plastics and implementing purchasing policy to support these products. With the right policy settings in place manufacturers will proactively seek to engage designers with a demonstrable commitment to sustainability.

e. Price shocks

Global events are causing significant fluctuations in the price of fossil fuels causing significant disruption to the petrochemical industry.

i. Covid-19

Covid-19 initially caused a reduction in demand for fossil fuels as many communities went into lockdown to slow the spread of the virus. The price of oil fell from around $60 per barrel at the start of 2020, to a low of about $11 in April, stabilizing at around $40 for the remainder of 2020, as the market adapted to decreased global demand (Macrotrends, n.d.). This decrease in the price of oil resulted in the price of virgin plastics (including the monomers used to make them) falling, in some cases below the cost of recycled materials. Some plastic markets have since recovered, initially buoyed by increased demand for PPE and a surge in demand for packaging caused by the rapid increases in online sales and home delivered food.[10] Conversely, demand was affected by the cancellation of sports events and concerts, reducing requirements for SUPs (J. Poole, 2020).

Decreased oil prices also impacted the financial models driving investment in gas-fired cracking plants across the United States. A major global transformation away from reliance on oil as the major source of plastics (primarily manufactured in Asia) toward cheaper gas from fracking activities (primarily in the United States, where nearly half a million wells using fracking technology were active by 2019) was disrupted by Covid-19. The lower oil price impacted the business plans upon which capital raising for fracking investment. According to research firm S&P Global Platts, capacity to produce ethylene from gas rose 41 per cent on the Gulf Coast over the five years to 2020 while margins dropped more than three-quarters over that period. The Gulf Coast ethylene margin dropped to $127 a metric ton in 2020 from a high of $558 in 2015. In response, Saudi Aramco,

LyondellBasell Industries and Chevron Phillips delayed plans for three Gulf Coast investments totalling $17 billion (Chaudhuri & Eaton, 2020). Four additional petrochemical facilities planned for the same area were placed on hold indefinitely as the appetite to finance these projects evaporated (Marusic, 2021).

Covid-19 caused turmoil in energy markets: disrupting the long-term investments plans of petrochemical organizations and creating significant disruption to the landscape of the social-technical regime in which it operates. Covid-19 shook the system and demanded radical reorientations within the established regime. As the world began to adjust post-pandemic a new threat to global energy markets emerged.

ii. War in Europe

Conflict between Russia and Ukraine escalated significantly following the full-scale invasion of Ukraine on 24 February 2022. The ongoing war has resulted in increased turmoil to energy markets, with gas prices in particular increasing, as Russia restricted supplies to Europe. Dutch TTF Gas, a leading European benchmark price, reached over €320/MWh in August 2022, more than four times the price at the start of the year. Oil prices have also remained elevated, particularly since OPEC members cut production.

On the positive side, souring relations with Russia and elevated prices for fossil fuels are accelerating the shift to renewables. Germany has committed to install sufficient sun and wind power facilities to generate 80 per cent of the country's gross electricity use by 2030, double the share in 2021 (Jordans, 2022). Price and supply fluctuations are increasing risk for investors in fossil fuel dependent projects. In another encouraging development, activist organizations are leveraging energy scarcity in Europe to demand that fossil fuels should be diverted away from the production of virgin plastic to help relieve scarcity in domestic energy markets. In 2020, plastic production was responsible for nearly 9 per cent of the EU's consumption of gas and oil (CIEL and Break Free from Plastics, 2022).

3. Summary

In the real world, changes at the landscape level are not confined to those presented above. Covid-19 has resulted in many countries borrowing to support their economies through increased government spending. As they strive to reduce these debts, governments are more likely to consider implementing levies and taxes (including carbon or plastic taxes) to raise revenue. After the oil crisis of 1973, governments took the opportunity to significantly increase the taxation on oil to help resuscitate their debt laden economies (Carbon Tracker Initiative, 2020, p. 34). If history repeats, then government response could foster a supportive environment for the growth of the renewable plastics industry.

While I have attempted to itemize significant landscape disruptions, they do not operate independently but intertwine, with outcomes determined by combined effects. Governments will prioritize incentives and investments dependent on local conditions and influenced by the intensity of lobbying activities. Policy is directed by public opinion, which is shaped by the media and the effectiveness of campaigns from NGOs.

The MLP model recognizes landscape disruptions can act as catalysts to challenge the status quo, creating an opportunity for new, more sustainable technologies to breakthrough. In the face of significant disruptions, actors are driven to seek creative solutions, adapting to new market conditions and experimenting with challenger technologies. Previous shocks to the industry, like the 1970s oil crises, were primarily driven by sudden price changes, from which the entrenched regime quickly recovered, stabilizing the system. In the current situation, price shocks are being compounded by a range of factors, including growing concern around climate change, the environmental impacts of plastic and a growing awareness of the plastic waste problem, thanks to China's National Sword. The long-term impact of Covid-19 and war in Ukraine on the fossil plastic market is still being played out. It is not possible to predict whether the combined effect of these exogenous events will significantly destabilize the entrenched regime and force it to adapt.

MLP also cautions that to succeed in taking advantage of these temporary disruptions, niche innovations need to be developed and sufficiently mature to take advantage of the opportunity (F. Geels, 2007, p. 405). Renewable carbon materials are more advanced than they were in the 1920s or 1980s and more suited to replacing virgin plastics across many applications. Despite this notable progress, the widespread adoption of renewable plastics will encounter numerous impediments. The following chapter scrutinizes these barriers to enable the development of strategies to support designers and manufacturers who seek to assume an assertive role in catalysing the shift away from virgin plastics.

7 Specifying renewable carbon-based plastics

For renewable plastics to be widely adopted they must first gain support from a range of actors, including designers and manufacturers. This chapter examines the factors influencing the choice of materials for consumer goods and the obstacles that must be overcome by those seeking to promote the use of renewable plastics.

Extracts from interviews with designers and manufactures have been selected to illustrate the complex issues and decisions that require addressing when working with renewable plastics. These real-life experiences of experimentation with renewable plastics are examined to advance the scaling-up and scaling-out of the use of these materials. While case studies focus on plastic chairs many of the same challenges will confront all those seeking to promote the use of renewable plastics in other consumer goods.

In Chapter 3 a high-level vision toward a more sustainable plastics industry was outlined, but practitioners need guidance to follow a pathway toward that goal. At the product level, which combinations of materials and manufacturing technologies deliver more sustainable outcomes? More specifically, without evaluating the full life-cycle impact of products made from recycled plastics and bioplastics how can designers or manufacturers know which combinations of materials and manufacturing technologies are less harmful to the environment?

Without this information, system-level theories are of little use to individual designers or manufacturers who feel the moral responsibility to develop more sustainable products but are faced with frequent, pressing design decisions. This is, perhaps, best exemplified by the response Philippe Starck gave when I asked him for his thoughts on recent theories (Design for Sustainability Transitions or Transition Design) in an interview conducted for this study:

This is a joke. It's really, because there is no theory for me of this type. We just have to do our best without any rules or theories. It is a fight; it is a fight for ecology, for less materiality and more affordable prices. The duty of the designer – and the manufacturer – is to work for ecology and energy saving, while the duty of the consumer is to buy these responsible products. That is all. There is nothing more complicated.

Starck's comments reflect an emphasis on technocentric solutionism, the belief that social problems can be solved by technology. While this perspective may seem outdated to some, it is indicative of the practical priorities of working designers faced with making decisions daily. Design is practice-based, while transitions are theory-based. Theories and concepts of system innovation are too complex and deemed not useful by those who design and develop products (Gaziulusoy, 2010, p. 45). While practitioners may be motivated to contribute to a more sustainable future, they want and need practical guidance on where to transition to and how to get there. This section aims to begin to address these needs by examining the barriers that must be overcome to promote the use of renewable carbon plastics in design.

Figure 7.1 captures the many factors that must be considered while designing a product, with specific reference to a chair. Six major stakeholders are shown (in caps) and their roles in the decision-making process are indicated by the positioning of the decision labels. The roles of the manufacturer have been highlighted, as they usually ultimately decide material choice – the focus of this chapter. Manufacturers can be expected to take some decisions without consulting with stakeholders. For instance, the choice of energy provider is likely to be the sole responsibility of the manufacturer, as they are financially accountable for this decision. For many other decisions consultation with designers, retailers and/or material manufacturers/engineers are likely to occur before decisions are finalized and these are indicated by the placement of decision labels. Final choice of material is usually decided by the product manufacturer, but they most likely seek input from the designer and advice from material specialists before making their final decision. The designer has an opportunity to influence this important decision at least in some instances, as discussed in Chapter 3.

Many of the decisions illustrated are not mutually exclusive. The cost of production will be partly determined by the materials chosen and the manufacturing technology employed. End-of-life impacts can be determined by other decisions or specified as an input to those choices. The choice of materials is central to

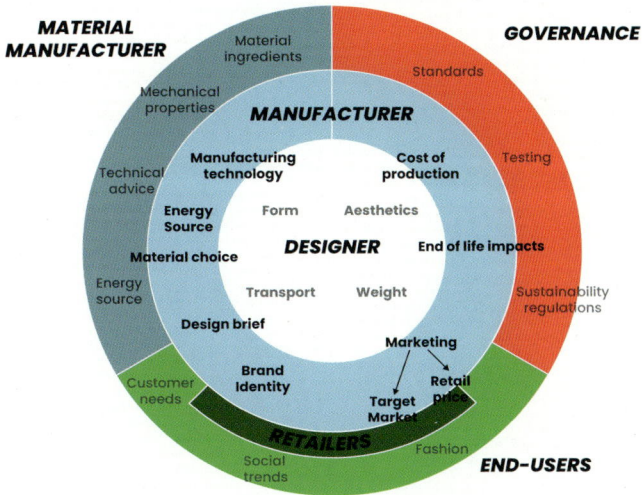

Figure 7.1 Chair development – sphere of influence. The decisions that need to be taken when developing a design are shown with those influenced by the manufacturer highlighted in bold. Where decisions are impacted by other actors (labelled in dark blue caps), this is indicated by the positioning of the decision labels.

assessing the environmental impact of a plastic chair (or most other products that do not consume resources during use). Material choice is a potentially complex decision, especially where new or unfamiliar materials are being considered. First, those involved with the decision must access reliable and up-to-date information on materials, their availability and cost. Designers and manufacturers need to be confident that the material's physical properties are appropriate for the task. Manufacturers must satisfy themselves that the required means of production and skilled labour needed to work with the material are available. The complexity of those factors can act as barriers, deterring the selection of renewable plastics and stymying change.

The barriers confronting those who seek to incorporate renewable plastics into their product designs can be summarized under three headings: awareness, availability and lock-in (Figure 7.2). Awareness refers to the lack of knowledge and understanding about renewable plastics, their properties and their potential applications. Availability refers to the limited supply and lack of commercial scale production of renewable plastics. Lock-in refers to the resistance to change due to the established use of traditional plastics and the sunk costs associated with existing production processes and supply chains. By understanding these barriers

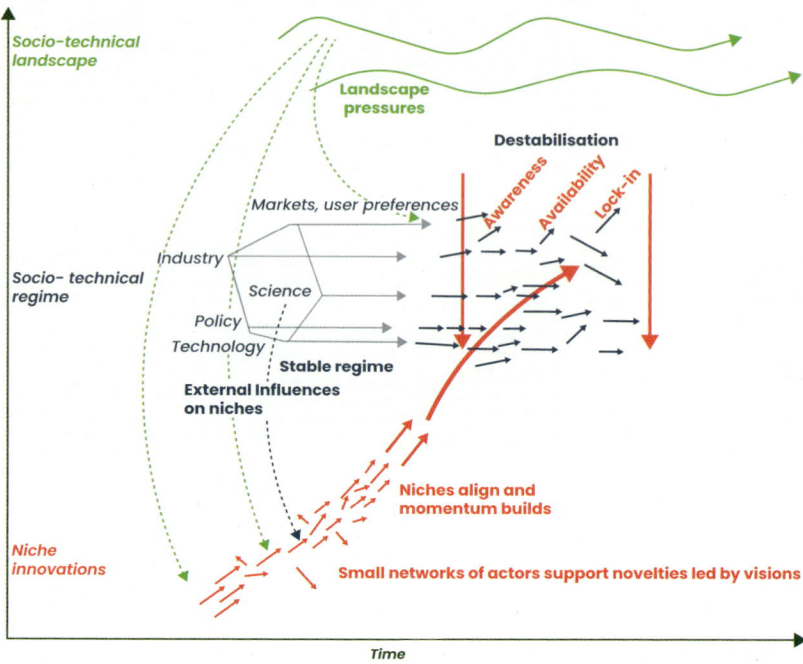

Figure 7.2 Issues resisting challengers to the established socio-technical regime. Adapted from Schot & Geels (2008).

and the experiences of those who have overcome them, designers and manufactures are better prepared to advocate for the adoption of renewable plastics. At the end of the chapter the relevance of a strategy recommended by MLP theory to accelerate the update of a new technology is considered. This analysis highlights the knowledge gaps that need to be addressed to encourage manufacturers to explore the potential of renewable plastics. Designers can act as agents of change by educating, supporting and advocating for renewable plastics.

1. Awareness and knowledge

Rapid advances in renewable plastics, particularly bioplastics, make it challenging for designers and manufacturers (and their educators) to keep informed of developments. Designers often limit their selection of materials to those they are familiar with, and this can often be attributed to a lack of awareness of alternatives (Anssary, 2006). Awareness, then, is the first barrier to overcome for the introduction of any new material. For designers and manufacturers to work with new materials they need to become knowledgeable about their physical properties and processing requirements. Developing or obtaining reliable, independent, unbiased information about new renewable plastics is particularly challenging. Information on the environmental impacts of materials if often simply unavailable (not yet gathered), too costly to obtain or has not been independently verified (Waage, 2007, p. 648). Confusing and conflicting messaging circulating in the market is made worse by deliberate misinformation campaigns and negative storylines initiated by incumbents.

Professional associations can often be expected to take the lead in promoting new technologies through journals, conferences or dedicated workshops (F. Geels & Deuten, 2006). However, despite the number of associations and publications servicing the design industry, participants in this research failed to identify common sources of information on materials. Designers often told me that they discovered new materials by chance. Other designers, for example, Sarah Gibson, co-owner of DesignByThem, reported using search engines, blogs and industry contacts when seeking new materials. To date there are no established, trusted sources of information for designers seeking to work with renewable plastics.

British metallurgical engineer Michael Ashby, working with design consultant Kara Johnson, reported similar findings. While engineers had ready access to information about new materials and shaping techniques, they found industrial designers to be frustrated by the fact they did not have access to equivalent support (Ashby & Johnson, 2003, p. 25). Alaa El Anssary conducted research on German and Egyptian product designers and engineers, discovering a lack of information about new material or techniques were common among designers. While engineers could access the information they needed, designers often found the available information to be presented in a complex, scientific fashion, too detailed for their needs (Anssary, 2006, p. 184). Anssary also reported the absence of a central, trusted source of information among the designers he interviewed.[1]

Dan Armstrong summarized additional challenges facing designers trying to get the information they need in a globalized market:

> If the client [asked] can I make it totally environmentally friendly, whatever that means. [We would say], 'Let's look into that.' Then you go to

China and talk to them about it and they will always give you an option and say, 'Yes, no worries, we have this.' It's just a business relationship and keeping people happy. But the truth, the real truth of those materials and the end results I'm not sure of. No one really knows. People want just desperately to put an eco-sign somewhere on their packaging and go, 'Is it all good? Sure? Cool, let's go let's make some money.'

Armstrong's experience illustrates the challenges faced by many designers operating in a world where production is increasingly outsourced around the globe, introducing another layer of complexity to their attempts to get accurate information about materials.

a. Intellectual property

One of the biggest challenges facing those who seek to work with renewable plastics is material manufacturers' secrecy about their products. Their reluctance to release details of ingredients and/or additives prohibits a detailed assessment of environmental impacts and can create uncertainty around the physical properties of the material. Barber and Osgerby reported they were unaware of the detailed composition of the recycled plastic used for their *On & On Stacking Chair*, manufactured by Emeco (Figure 5.24). Having converted manual sketches into 3D models these were then shared with the engineering team at Emeco, who were relied upon to ensure the structural integrity of the design as discussed in Chapter 6. In this case, separation of responsibility was crucial in giving the designers confidence to experiment with the new material. For designers operating without the backing of a committed manufacturer those unknowns are a formidable barrier to experimenting with unfamiliar recycled plastics.

My interviews with designers and industry representatives highlight the difficulty in obtaining the information needed to assess the environmental impact of the materials with which they are working. Making the 'best' sustainable choice requires transparent access to detailed information on the material sources, ingredients and supply chains (Despeisse et al., 2017). In the rapidly developing and competitive bioplastic market, material manufacturers seek to protect their investments and remain guarded around revealing details of ingredients; in efforts to prevent competitors taking advantage of their R&D.

In some cases, this lack of transparency can lead to designers inadvertently using bio-based or hybridized products. This was the case for industrial designer

Ander Lizaso's project for Alki, a manufacturer committed to sustainability principles (Figure 5.12). As the *Kuskoa Bi* project progressed it emerged that the biomaterial they were using was a hybridized product with a percentage of fossil-based plastics mixed with the bioplastic:

> There is a percentage that is fossil fuel based … I don't know if the other part of the percentage is PP, I don't think it is … When we approached the people, we work with for this material they were quite opaque, it was kind of a guess …

Finding out the source used to create the bioplastic component of the material proved equally difficult:

> We had a very wide, very broad definition of the material and we didn't know if it came from corn starch or sugar cane or even beetroot and we also were quite conscious about the implication of each of them. The marketing [department] went for corn because it was really visual. But in the studio, we were hoping that it wasn't corn because the implication is that corn shouldn't be used for making chairs. We know [now] that it comes from sugar cane [stover].

Lizaso's experience illustrates that designers working with bioplastics face additional challenges when attempting to evaluate the environmental impact of their work. Lizaso was not alone here; Rashid reported he has been unable to identify the exact contents of the 'bioplastic' used for his *Siamese Chair* as this information was not available to the furniture manufacturer (Figure 0.3). Even when material suppliers are prepared to share information, they might source organic ingredients from different suppliers (depending on season and availability), which could impact any attempt to measure the sustainability of their outputs. When sourcing different crops, the use of genetically modified organisms (GMOs) together with the use of pesticides and fertilizers are likely to vary. Transportation methods used and distances travelled to the processing plant will invariably change with the feedstock, with varying downstream impacts on the overall environmental performance when bioplastics are used in products.[2]

In the absence of full disclosure on the contents of materials, prospective product manufacturers are unable to complete detailed life-cycle assessments or any detailed level of eco-audit of their products. While it is understandable that organizations investing in the development of new materials want to protect their intellectual property, this secrecy also impedes diffusion of new materials.

b. Recycled plastics

Recycled plastics have been available for many years, allowing designers and manufacturers to become acquainted with them. Despite that, the designers and manufacturers I interviewed often expressed significant confusion about those materials, particularly the energy consumed during their production and the impact on their environmental performance. Karim Rashid commented:

> Most people think plastics become cool when they are recycled. People like this thing of taking plastic out of the sea and creating something out of it – to be honest all the machinery, infrastructure transport – and the whole recycling process to first of all find that plastic, move it somewhere else, move it to a few different places, then process it is probably more polluting than creating something with plastic in the first place.

Rashid was not alone in that observation, with designer and academic Manuel Garcia claiming that 'in most cases it [recycled plastic] is actually a marketing tool – people don't evaluate how much you are saving in energy consumption and carbon footprint because of your method – they just look at the material'.

Incorporating recycled materials into a design does not necessarily guarantee better environmental outcomes. How the material is collected and sorted by type, the amount (and source) of energy consumed during the recycling process and the end-of-life options for the material are among the significant factors impacting the environmental profile of a design. Post-consumer recycling is particularly impacted by the difficulty of accurately identifying and sorting plastics from combined waste streams.

Recycling has long been promoted by the plastics industry as the solution for the plastic waste crisis. Even though only 9 per cent of all the plastics ever produced has been recycled, and with global recycling markets in turmoil, the petrochemical industry continues to promote recycling as the solution to reducing the industry's environmental impact (Geyer et al., 2017b, p. 1). Fossil plastic is often portrayed as an ideal candidate for the circular economy, infinitely recyclable, the ultimate cradle-to cradle material, it can be used without guilt. S.CAB claim that chairs made from GoGreen recycled plastic can be recycled 'infinite times in a virtuous process aimed at not introducing new plastic into the environment' (S-CAB, n.d.).

The claim that plastics can be continuously recycled using existing mechanical processes is, however, a myth that maintains a strong hold in the design community. Several designers told me that they believed that plastics can be recycled infinitely, without loss of performance. Thomas Pedersen (Danish designer) claimed that if plastic is in a closed circle, then it can be used forever. UK-based design team Barber Osgerby talking about their *On & On Stacking Chair* said: 'for us the important thing was that it is recycled plastic that can be infinitely recycled without downgrading, which is brilliant'. Garcia claimed PP can be recycled forever and went on to claim that PET can be infinitely recycled without losing mechanical properties. Gibson commented:

> We've always thought that recycled plastics – HDPE – can be used over and over again without any deterioration, so I have always had this strong trust that they will be used again.

In fact, HDPE (one of the easiest plastics to sort and recycle) experiences thermo-mechanical degradation during processing, which alters its structure depending on the temperature and stress exerted during the mechanical recycling process (Loultcheva et al., 1997, p. 77). Similarly, all plastics are impacted by the heating and processing treatments involved with mechanical recycling, weakening the links between the polymer chains.[3] Reprocessing recycled polymers is always accompanied with degradation resulting from molecular chain scissoring, branching and crosslinking (Yin et al., 2015, p. 2899). Even the founder of the Vinyl Institute conceded in 1989 that 'recycling cannot go on indefinitely, and does not solve the solid waste problem' (Centre for Climate Integrity, 2024, p. 9). Reuse strategies can only offer part of the solution toward a more sustainable system as continuous population and economic growth mean that even in a perfect recycling regime, where 100 per cent of material is recycled, virgin material would still need to be added to maintain the mechanical characteristics of the material and meet ever-increasing demand while compensating for losses in the recycling process (Bjørn & Hauschild, 2013, p. 329).

Melt blending can help overcome these issues, mixing recyclates with virgin material creates a composite which displays performance similar to virgin plastic. However, a UK paper summarizing the challenges and opportunities of plastic recycling found that it is usually not technically feasible to add recovered plastic to virgin feedstock without decreasing at least some quality attributes of the virgin plastic such as colour, clarity or mechanical properties, like impact strength (Hopewell et al., 2009, p. 2119). Most current recycling activities can be more accurately described as downcycling. Recycled HDPE is commonly used for applications such as milk crates, bins, plastic lumber, underground pipes or

Figure 7.3 Envier outdoor furniture 'constructed from 85 per cent recycled plastic and at the end of its life is fully recyclable and can be repurposed over and over'. Courtesy Integrated Recycling.

plant pots. PP is usually recycled only once and mixed with 50 per cent virgin material to make new products, including outdoor furniture (Figure 7.3) and playground equipment.

While PET can be infinitely mechanically recycled, producers add antioxidants or blue colourants to counter discolouration (yellowing), caused by other additives or contaminants during the recycling process. Multiple heating cycles can cause the breakdown of long polymer chains rectified by the addition of additives to improve the flow rate of the material (Hopewell et al., 2009). PET can only be indefinitely mechanically recycled with the addition of more chemicals. In addition, researchers are beginning to investigate the chemicals contamination of rPET with some alarming findings. One study showed that antimony and BPA, both of which can cause chronic health problems including cancer, were among 150 food contact chemicals found in rPET (Gerassimidou et al., 2022). The authors conclude:

> It is important to note that the exact chemical composition of rPET is essentially not known unless measured. Certain hazardous FCCs may or may not be present in rPET on the market and only a case-by-case analysis can establish the presence, levels and safety risks of potential contaminants of concern.
>
> (Gerassimidou et al., 2022)

The petrochemical industry is now promoting chemical recycling as a solution that ameliorates the degradation caused by mechanical recycling. The term 'chemical recycling' (or the more palatable 'advanced recycling', as the petrochemical industry is attempting to rebrand it; it is also sometimes called feedstock or molecular recycling) includes several technologies (pyrolysis, gasification and depolymerization) that break down the molecular chain of polymers into liquid or gases which can then be processed into constituent chemicals (Francis, 2016, p. 59).[4] By breaking down plastics to their components they can theoretically be reborn, displaying the same properties as their virgin parents. Industry representatives promote this technology as 'true circular recycling', often ignoring the fact that significant resources will undoubtedly be lost during the energy intensive processes (Latham, 2021).

Most existing chemical recycling plants use heat and pressure to produce pyrolysis oil that, when burnt as fuel, release GHGs and exit the circular economy (Oladejo & Rollinson, 2020, p. 14).[5] Critics argue that chemically transforming plastic into pyrolysis oil is not recycling; it is just another way to create more fossil fuel to burn and can more accurately be described as waste-to-energy (Break Free From Plastic, 2021). Outputs are recovered for energy production rather than recycled. Indeed, as of March 2023, at least four of the states in the USA have already legislated that chemical recycling cannot count toward recycling and waste reduction targets (GAIA, 2022).[6] Proponents highlight that any technology that displaces consumption of virgin fossil oil offers a net gain for the environment.

In 2023, the Plastics Industry Association presented research to the Federal Trade Commission concluding that consumers are not concerned about how plastics are recycled and that a large majority of Americans support advanced recycling and agree that it should be considered recycling (Plastics Industry Association, 2023). Although the definition of advanced recycling presented in the question posed to respondents was arguably biased: 'a method used to address waste, used plastics can be broken down into smaller molecules that can be used to make new products, including new plastics. There is no burning of plastics involved in these processes. Do you believe this is an example of recycling or not?' Given the wording of the question it is perhaps unsurprising that this industry association could report such high levels of support for these new technologies!

Despite industry enthusiasm, chemical recycling accounts for less than 1 per cent of installed recycling capacity globally (European Investment Bank, 2023, p. 39). These activities can then be seen as an incremental response, a typical defence manoeuvre by an incumbent regime which in this case attempt to exert control over the recycling industry, previously led by the waste regime.[6] As the petrochemical regime exerts control over the industry there is a danger that small-scale operators from the waste industry may find themselves locked out. For example, justifiable concerns with the quality and consistency of supplies may be used to legitimize decisions not to purchase products such as pyrolysis oils and gases produced by independent chemical recycling operators.

The scalability of 'advanced' technologies, mostly still in the experimental stage, their energy and resource efficiencies, together with the environmental and health impacts of toxins released during processes, all remain the subjects of debate by industry, academics and legislators (Biessey et al., 2023; Creadore & Castaldi, 2022; Oladejo & Rollinson, 2020, p. 21).

In July 2021, Greenpeace released a recorded conversation with a lobbyist working for Exxon, in which he admitted that if the company could not find solutions to the recycling problem it would at least find some talking points to raise with politicians as part of this lobbying effort:

> It's just like on climate change right so when climate change came, well it's here, but, well, when it started you started to have conversations to say, well you can't completely change the electric grid from coal and gas into wind and here's why.
>
> It's the same conversation, you can't ban plastics because here's why or you can't recycle, you know, legislate 100 per cent recycling, because here's why …
>
> We're doing the research, we're looking at our markets, we're looking at the chemistry and we're hoping to be able to come up with, if not solutions at least some reasons and some talking points to have with members of Congress.
>
> (Carter, 2021)

Greenpeace claim this to be the first time a representative from the energy industry has been documented admitting participation in strategies deliberately designed to delay or avoid legislation to lessen production of fossil plastics.

Sufficient information to enable a detailed assessment of the environmental impacts caused by chemical recycling is often unavailable. For instance, it is often claimed that some chemical recycling methods eliminate the need and cost of sorting plastics by type, ignoring the fact that plastics often still need to be separated from contaminants collected from mixed household recycling bins. For instance, pyrolysis requires inputs of at least 85 per cent pure polyolefins (Alliance to End

Plastic Waste, 2022). Other advanced recycling processes often still require plastics to be sorted and cleaned before processing (CSIRO, 2021b, p. 40). In a more specific example, a BBC article announcing a chemical recycling facility being built in the UK claimed that 'with a conversion rate of more than ninety-nine per cent, nearly all the plastic turns into a useful product' (Latham, 2021). The operator went on to make the unchallenged claim: 'the hydrocarbon element of the feedstock will be converted into new, stable hydrocarbon products for use in the manufacture of new plastics and other chemicals'. Exactly how much energy is consumed in this process, what the 'other chemicals' are, what (if any) additional processes are needed to complete the manufacture of these products and for what purposes they will be used remain unspecified. As chemical recycling becomes more prevalent, designers and manufacturers need to stay informed about these debates and question the detail of sustainability claims made by material suppliers.

As the quantity of recycled plastics entering the market increases, it will become increasingly necessary for manufacturers and designers to interrogate suppliers to establish both the source of the recyclates and the environmental impacts caused by their reprocessing and/or hybridization with virgin feedstock. In a truly circular system, plastics will be recycled multiple times and designers will need to become more knowledgeable about the provenance of their materials and the potential impact of reprocessing on mechanical properties. Using post-industrial waste from production facilities using virgin plastic provides surety on provenance and offers a short cut to secure quality feedstock.

i. Physical properties

Working with recycled plastics offer the advantage that many of the materials have a well-established track record, meaning their physical properties are well understood. Although the materials might not have been used for a similar project, their previous applications can provide some insights to their suitability for the intended task. Gibson used sheets of recycled HDPE, a material developed for outdoor use as a sound barrier at construction sites, for the *Butter* series (Figure 7.4), which she designed with Nicholas Karlovasitis:

> I guess we just looked at the context it was being used that thought that it would be appropriate for furniture and similarly with the Confetti range [made from recycled post-consumer and factory waste plastic], that material is put in the ground all the time – used for pipelines so probably sufficiently resilient.

Figure 7.4 *Butter Stool*, Nicholas Karlovasitis & Sarah Gibson for DesignByThem, 2011. Courtesy Pete Daly for DesignByThem.

Knowledge of this history reassured the designers that the mechanical properties of the material were up to the task.

Using recycled PP typically comes at the price of increased part defects caused by flashing, sink marks and short shots (Milliken, 2020). Processors can compensate by using a high ratio of virgin material or by running at higher temperatures to improve the melt flow to fill out their moulds, but this solution consumes more energy and can degrade the plastic and affect its mechanical properties. As a result, expensive modifiers are needed to compensate. Petrochemical companies are responding to the increased demand for recyclates and investing to address these shortcomings. Chemical company Milliken sells an additive designed to enhance the properties of recycled PP, eliminating the need to mix it with virgin feedstock. However, while reductions in both energy consumption and processing time are promoted (and undoubtedly appreciated by clients), the precise ingredients of this additive and their environmental impacts remain unknown (Milliken, 2021).

Shrinkage is also a potential challenge when working with recycled plastics. Pedersen reported that the shells for his *Falk* chair, made from post-consumer PP, were too open when first removed from the mould (Figure 7.5). At the time of our interview, Pedersen was working closely with the manufacture (Houe) to resolve the issue. A tool was being developed for use by the robot tasked with operating the injection-moulding machine. To contain costs the process needed to be optimized to ensure it could be executed within the existing three-minute cycle downtime. By forcing the shell closed and holding in place for a few seconds the issue could be rectified. Solving such problems after the project has entered production can have significant consequences, both financial and in terms of GHG emissions.

Figure 7.5 *Falk*, Thomas Pedersen for Houe, 2019. Robotic arm removing a *Falk* chair shell from the mould. Courtesy Houe.

Pedersen's experience is not unique. LEGO® abandoned highly publicized plans to switch to rPET from ABS when, after two years of testing, the company calculated the retooling required to accommodate the new material would result in higher carbon emissions over the product's life cycle (Milne, 2023).

c. Bioplastics

Bioplastics that are made entirely from renewable organic biomass offer·the potential to transform the reputation of plastics, from being viewed as an environmental criminal to a material that is valued as derived from and returning to nature. However, conflicting priorities among the actors involved with the development and use of bioplastics add to the complexities involved with understanding this material, as summarized by material scientist Sean Ferguson:

> Petrochemical companies want materials that integrate into existing processes, agribusiness want to promote heavily subsidised crops, environmental groups push for fossil-based alternatives, without fully understanding the impacts of these new materials.
>
> (Ferguson, 2012, p. 18)

Some of the designers participating in this research remain confused about the feedstocks used to create bioplastics and their properties. Despite the fact industrial designer Louis Durot works in research for a conventional plastic manufacturer, his opinions on bioplastics are outdated. He believed, for example, that bioplastics are not suitable for developing innovative shapes. He argued there would be supply problems if bioplastics became widely adopted as their manufacture consumes food crops. Another French designer, Eugeni Quitllet, held similar views, stating:

> But [with] bioplastic there was this ethical thing of saying well we are using corn which is food for some people we are using lots of water to grow the corn we're using process to make this bioplastic which are not 100 per cent ethical.

While first-generation bioplastics are made from food crops, subsequent generations can be made from non-food crops or stover (second-generation), algae or bacteria (third generation, although some definitions also include stover in this category) and even CO_2 (fourth generation). Subsequent generations do not, therefore, compete with agricultural land needed for food production, alleviating concerns about food supply.

With global population expected to approach 10 billion by the middle of the century any activities that divert already-scarce agricultural land away from food production will be challenged. Current demand for bioplastics is met with negligible impact on land use – estimated at only 0.015 per cent of the global agriculture area (European Bioplastics, 2020). If production of all plastics switched to biomass the amount of land needed is subject to much debate.[7] The debate is likely to become irrelevant well before it concludes, with higher generation feedstocks (such as, algae, bacteria and CO_2), which do not compete with food crops, becoming more prevalent in bioplastic production. In the short term, opponents highlight the use of agricultural land and associated impacts of deforestation, water pollution, soil degradation and loss of biodiversity are negative impacts of bioplastics, creating uncertainty among policy makers and potential customers. Further, while the industry distinguishes between first and subsequent generations of feedstocks it is unlikely that the distinction is understood in the public discourse, representing another communication challenge for those working with these materials. It is incumbent on designers and manufactures to satisfy themselves that their chosen bioplastics are sourced and produced ethically.

The varying sources of raw materials and methods used to produce bioplastics result in a wide range of materials with differences in the quality of materials produced. Unlike oil refineries which are relatively homogeneous, individual bio-refineries differ depending on the availability of raw materials (cane, beet or maize, for example) and the conversion technology used.[8] Additives are often used to improve the properties of bioplastics, which may impact their biodegradability or result in ecotoxicity effects.[9] For instance, per- and polyfluoroalkyl substances (PFAS) commonly referred to as 'forever chemicals' are often used as a moisture and grease resistant coating on the most commonly available bioplastic, polylactic acid (PLA), despite known health and environmental concerns.[10] All these variables present another barrier to adoption, as they cause uncertainty around securing homogeneous supplies with identical physical properties (Schot & Geels, 2008, p. 544). As we have already seen (Chapter 3) bioplastics are expected to remain a niche market for the foreseeable future and this is consistent with the findings of an analysis of transformations in the electricity, transport and heat markets in the UK which concluded that radical techno-economic reconfigurations struggle to diffuse, particularly without policy intervention (F. W. Geels & Turnheim, 2022, p. 301).

i. Physical properties

When considering bioplastics, the material's properties must be thoroughly examined as these can differ significantly from the fossil-based substances they are designed to replace. This can impact production, as manufacturing processes must be adapted to suit the characteristics of the bioplastic, a time-consuming and expensive exercise. Moreover, designers must be prepared to adjust their creations to suit the properties and characteristic of the bioplastic, which can impact the final form and appearance.

When working with bioplastics, pioneering actors face the challenge of limited independently verifiable information about many of the new materials emerging onto the market and very few have undergone rigorous testing over an extended period. Armstrong summarized the issue:

> Things haven't been around long enough. Polypropylene and ABS have been around long enough, so you know what is going to happen. But that is, unfortunately, saying we are such a slow-moving vehicle, it is going to take long time because we don't trust anything that's new.

Designers and manufacturers working with virgin plastics benefit from decades of research and development, allowing them to confidently predict how the material is going to behave when processed using well-established manufacturing technologies.

Konstantin Grcic identified this lack of track record as the main inhibitor to him experimenting with bioplastics:

> My clients … do want to use new materials but … they have to have certain certificates or guarantees that you know they have gone through testing that the material will more or less look the same and perform the same after five years of exposure to UV light for example or [can withstand] the structural stress and so on.

To circumvent the limited availability of data and lack of real-world experience, some designers have resorted to less sophisticated techniques. Lizaso explained the process when selecting material for the *Kuskoa Bi* chair:

> The first thing we did was of course the mechanical testing that you should do, but also leaving a shell outdoors and we knew that we didn't have three years to look at it, but we exposed it to some extreme conditions. We dumped it in water and it's very rainy here. That was the process at that stage.

Similarly, Rashid developed a bioplastic chair for A Lot of Brasil, explaining in an interview for this study the surprisingly casual process he undertook to satisfy himself that the proposed material was suitable for the task.

> You know you just you do a little research. I mean you take a piece of the moulding and you see the strength of it, you know, you rub things on it. I do things I can take keys and scratch things and you kind of get a good feeling of its durability.

Of course, some experiments with materials are not immediately successful. Rashid went on to explain how he was forced to adjust his original concept for the *Siamese Chair* to accommodate inherent weakness of the material (Figure 7.6). As detailed in Chapter 6, the bioplastic was insufficiently strong to be used for the legs and the shell of the chair had to be thickened considerable, with both adjustments negatively impacting the environmental performance of the chair. Rashid's experiences, together with the examples shared by other designers during interviews for this study illustrate the commitment required from designers when experimenting with new biomaterials.

Commitment is an expensive commodity. Experimentation takes time and can consume significant resources, requiring substantial financial backing. This was the situation facing Barber Osgerby when I interviewed them in July 2019.

Figure 7.6 *Siamese Chair*, Karim Rashid for A Lot of Brasil, 2014. Rear view of the 9.8 kg chair showing the bulky shell needed to satisfy strength requirements. Courtesy A Lot of Brasil.

Referring to a bioplastic chair project they had been working on for four or five years with an unnamed manufacturer the designers reported the material turned out to not be sufficiently durable. Barber Osgerby had to work with the material manufacturer to re-engineer it:

> The problem is, with this eco-plastic, they are not brilliant in rain, not brilliant in extreme heat, they tend to warp a bit, which would be fine if it was a very low-cost product. But, if it is going to cost the same as a high-end product people are going to say, 'I can't afford to have my chair start moving in the sun'.

Barber Osgerby had plans to launch this bioplastic chair at the *Salone del Mobile* in April 2020. That exhibition was cancelled due to Covid-19. The *Alpina Chair*, featuring a bio-based back rest supported on a timber frame and seat, was eventually launched by Magis in 2022 (Figure 7.7). Separately, Magis reported experimenting with some (unnamed) renewable plastics but abandoned the projects following trials that failed to deliver the required results.

Gabriele Chiave, from Marcel Wanders' office, also reported working with Magis and Philippe Starck on a liquid bamboo pressed chair. The project was

Figure 7.7 *Alpina*, Barber Osgerby, 2022. Seat and legs in solid ash, back in bio-based polypropylene. Courtesy Magis.

abandoned after eighteen months after the chair repeatedly failed engineering tests. Poor thermomechanical performance and climatic resistance properties are often cited as deficient when evaluating bio-sourced polymers. In 2021, Vitra undertook a detailed evaluation of available bioplastics to assess their suitability for manufacturing long-lasting furniture components but found that none were up to the task (Kries et al., 2022, p. 163). All these experiences highlight the financial risks involved in attempting to work with renewable carbon materials. While these well-established actors are well placed to absorb unexpected costs, many others are unlikely to be able to take such risks.

Even when an appropriate bioplastic has been identified, experimentation is needed to examine its properties and optimize production processes. Victor Macadar, a client of Bambacore, an Australian supplier of bioplastics, discovered that the bamboo reinforced bioplastic he was using could be injection moulded at lower temperatures, saving energy and money. However, while the injection speed time is slower the cooling time is reduced, as the product can be removed from the mould more swiftly. Overall, the cycle time is not significantly impacted, but Macadar cautions that these factors can vary, with significant cost implications, as experiments must be undertaken in conjunction with the moulding company. Macadar highlighted during our interview that both designers and manufacturers need to understand that bioplastics can also have varying shrinkage rates when removed from the mould. Lizaso also experienced significant shrinkage of the shell for his *Kuskoa Bi* chair (Figure 5.12) which was only detected after the project had entered production, leading to inconsistencies in the end product, and necessitating a redesign of the base to accommodate the variance:

> You can have a shell that is almost two centimetres wider than the other because in the cooling process it [bio-based PP] moves quite a bit, much more than normal PP shells.

Designers working with bioplastics also reported that these materials can be more corrosive in the mould, potentially leading to higher wear and tear and associated costs. Changes to hot runners and even screw design may also be required. Moisture levels must be minimized during the injection-moulding process, or the end product might become brittle and weak. PLA offers a comparatively lower processing window and has a lower melting temperature while PHA also has a narrow processing window and a slower crystallization rate (Fachagentur Nachwachsende Rohstoffe e. V., 2016). Disappointingly, at least one of the designers interviewed for this study had abandoned bioplastics due to negative experiences. Garcia 3D printed *Voxel Chair v1.0* from PLA but found the material degraded over time and discoloured in sunlight (Figure 5.22). He has since moved away from using that material in favour of rPET. Poor experiences with early bioplastics made from first-generation feedstocks risks tarnishing the reputation of the sector before the industry has a chance to mature.

To mitigate the risks associated with working with bioplastics material manufacturers have developed bio-based materials, specifically designed to match the physical properties of the fossil plastics they replace more closely, thereby minimizing the adjustments needed to manufacturing processes. Often also referred to as bioplastics, these bio-based or bio-derived hybridized materials only contain a proportion of organic content, sometimes very low. Often the exact composition of bio-based materials is not disclosed, making it impossible to accurately calculate their environmental impacts. The absence of standards has allowed material manufacturers and marketers to create and exploit confusion in the market, often implying environmental benefits but offering little explanation. As a result, designers are struggling to understand the environmental impact of these mixed materials, as Armstrong explained:

> The promoters of bioplastic sometimes have just 20 per cent organic content and the rest is just traditional plastic and then, of course, at the end of life is crazy … That's my biggest problem – you've really spruiked this, you made your brochures all look eco and there's a rainforest on the front cover, that's nice and you when you look into page one you see what you've done you are not really there yet.

Mixing biological-based and fossil-based plastics diminishes end-of life prospects, as these bio-based materials cannot be recycled and will not biodegrade. These hybridized materials should not be confused with true 'drop-in' bio-based materials such as bio-PE, which are made entirely from renewable organic materials and sometimes referred to as biopolymers. Materials such as Braskem's 'I'm Green' bio-based PE is derived from sugar cane and can be mechanically recycled together with their conventional PE, as they possess the same chemical structures, despite the difference in feedstock.

ii. End-of-life prospects

The term 'biodegradable' lacks a standard definition for the time it takes to decompose. All substances will biodegrade eventually, although it might take a few hundred years. Some of my interview subjects were confused by biodegradability claims, with one designer stated that all bioplastics biodegrade, making them unsuitable for use in manufacturing consumer goods in their opinion. Currently 'degradable plastics' is defined to include oxo-degradation, thermal-degradation, photo-degradation, biodegradation and compostable plastics.[11] As an indication of how contested this space is even the definition of 'compostable' is still being debated. A German court ruled it is deceptive to market plastics as compostable as they do not convert into compost, arguing they convert 90 per cent into CO_2 gas without leaving any nutrients.[12] A Danish court ruled that plant pots made from a bioplastic (PLA) could not be described as biodegradable, because they only degrade in the special conditions found in industrial composting facilities and will not decompose, within a reasonable time frame, in a domestic compost environment.[13]

Oxo-biodegradable plastics are being promoted by the petrochemical regime as the solution to this problem. These conventional fossil-based plastics have been treated to breakdown when discarded in the natural environment. If these plastics actually break down at the molecular level or simply fragment into microplastics remains highly contested despite mounting scientific evidence to support the latter (Abdelmoez et al., 2021; Sciscione et al., 2023; Simona et al., 2020). The incumbent regime has a vested interested in promoting this technology, as it will allow them to continue to manufacture virgin fossil plastics, while the debate about the long-term environmental impacts of these materials continues.

Understandably, there is much confusion about the end-of-life prospects for bioplastics among designers and end-users alike, with evidence to demonstrate that consumers are particularly bewildered by conflicting claims.[14] Macadar told me that the biodegradation time is usually the first question asked by his perspective clients. Attempts to obtain a definitive answer from the material supplier on the life expectancy of the material were not successful. Macadar called for more regulation or certification, specifying how long materials take to degrade and under what conditions, to better inform the market and provide clarity for end-users.

Bioplastics create complexity for recyclers, designers and manufacturers interested in understanding the end-of-life prospects for products. Ideally bioplastics need to be sorted from fossil plastics in the waste stream, but that is practically impossible. The RIC labelling scheme has not yet been extended to cover bioplastics, making it challenging to identify the material, even by manual inspection. To meet the high-speed sorting requirement of material recovery facilities sophisticated sorting equipment is needed. This investment is unlikely to occur until bioplastics account for a significant proportion of the material entering recovery facilities. Meanwhile bioplastics cause significant problems for recyclers and can cause enormous environmental damage. A very small level of contamination from bioplastics can cause entire batches of recyclates to be sent to landfill or incinerated.[15]

Even if biodegradable bioplastics are separated from the waste stream and sent to industrial composting facilities it is likely they will be rejected, due to the complexity of identifying and sorting appropriate feedstock. Refusing all plastic waste is a simpler option. Additionally, biodegradable bioplastics often take longer than organic waste to fully decompose, adding costs and complexity for compositing operators (Körner et al., 2005, p. 409). As bioplastics become more prevalent the challenges around identifying, sorting and processing them will intensify, potentially driving up waste processing costs as additional investment in sorting technologies will be required.[16] The OECD warns these trends could make investment in waste-to-energy facilities an attractive solution for all plastic waste, locking in a less preferable solution to recycling from an environmental perspective

(OECD, 2013, p. 52). Designers and manufacturers working with bioplastics need to understand and clearly articulate the preferred end-of-life treatment for their products in communications with end-users and waste processors.

iii. Limits on growth

For designers and manufacturers interested in exploring renewable plastics, the confusion caused by a lack of independently verifiable information about the properties and potential of new materials has slowed the adoption of bioplastics. In our interview, Gregg Buchbinder, CEO of Emeco, told me that despite their well-publicized commitments to sustainability, they preferred to adopt a wait-and-see approach, believing it is too early to experiment with bioplastics. Buchbinder specifically cited 'misleading or glorified information' as a reason for the company's hesitancy adding, 'we always want to be sure we are cognisant before putting in the time and effort necessary to work with a new material'.

John Tree, from Jasper Morrison's office, held similar views saying that he considered bioplastics: 'something to keep an eye on, as it is maturing very fast', but added 'we are at a point of inflection at the moment where things become possible and realistic'. Ron Arad also told me that he preferred to wait and see how well bioplastics performed in real life before working with them. Likewise, Bertjan Pot commented that he did not have the resources or interest in experimenting with new materials. His preference is to wait for the market to mature and 'then I know if I have something to add to it as a designer or an artist or not'.

It is challenging for those interested in working with bioplastics to develop a detailed and up-to-date understanding of a material's suitability for a specific project and its full life-cycle environmental impacts. With new materials entering the market, sourced from different generations of feedstocks, the debate around the merits of bioplastics is likely to become more complex and challenging. Adding more complexity is the fact many of these novel materials are being offered by relatively small, newly established market players, who lack a proven track record to validate their assertions.

2. Availability

For renewable plastics to be incorporated into the portfolio of materials used by product designers the materials need to be readily accessible, and their availability publicized. Bioplastics face a particular challenge in terms of availability. Macadar had to conduct a global search to find a material suitable for his product. He reported that as bioplastic producers are relatively scarce and often in start-up mode, they lack well-resourced marketing and communication departments, making them hard to locate. Recycled plastics are likely to become scarcer, as supply fails to keep up with the growth in demand caused by the shift away from virgin plastics in SUPs.

In a comparatively small and remote market, like Australia, the choice of renewable carbon materials is extremely limited, as Gibson discovered when surveying the market for recycled plastics during the first decade of this century. During an extensive search for suitable materials, Gibson only found recycled pellets suitable for injection moulding, and one other product, developed for roadside use as a sound barrier, later used for the *Butter* range. Since introducing the *Butter* range, the designer has been forced to look overseas and change the material supplier three times as the recycled plastics market struggled to establish. Referring to the *Fenster* range (released in 2020) Gibson added that despite extensive searching, she had failed to locate a sustainable alternative to acrylic glass used for the back and seat (Figure 7.8). The company decided, reluctantly, to release the product made from acrylic glass while continuing with attempts to source a more environmentally friendly substitute.

Gibson was not alone in reporting shortages of suitable recycled plastics. Even internationally celebrated designers with access to well-resourced manufacturers

Figure 7.8 *Fenster* range, Sarah Gibson & Nicholas Karlovasitis for DesignByThem, 2020. Courtesy Pete Daly for DesignByThem.

reported difficulties in locating renewable carbon materials. Starck first sketched a design for Emeco in 2001. Emeco wanted to avoid the use of virgin resin but failed to locate a suitable alternative. A chair was finally produced in 2012 (Figure 7.9), after Emeco worked with material scientists to develop a recycled PET/wood blend with sufficient strength for the task.[17] Emeco has continued to improve the performance of its recycled material, making the recyclate stronger and able to withstand harsher weather conditions, while increasing the proportion of recycled content.

Pedersen also reported difficulty in identifying a source for recycled household waste plastic for the shell of his *Falk* design. Local suppliers were unable to meet the accuracy of sorting required to develop the material. Early supplies of suitable material came from Germany, although, at the time of our interview (December 2019) Pedersen was confident that a Danish supplier could soon meet the necessary criteria and produce a suitable locally sourced material. Pedersen was eager to establish a local supplier as plastic waste is incinerated in Denmark to generate electricity.[18] By diverting this resource, Pedersen hoped to inspire other local designers to use the recyclate and raise public awareness of issues around waste-to-energy.

In addition to availability, the minimum order size offered by material manufacturers was also raised as an issue by some designers who reported difficulty in obtaining the modest quantities required for some projects. Mayda Diaz, from the Australian bioplastic agent Bambacore, were only able to supply materials in

Figure 7.9 *Broom Stacking Chair*, Philippe Starck for Emeco, 2012. Courtesy Emeco.

quantities of hundreds of kilos, which is prohibitive for those wanting to experiment with the material before making a large financial commitment.

a. Cost of materials

Another significant barrier for renewable plastics is that virgin plastics can be cheaper than recycled alternatives. During 2018, bioplastics cost two to three times more than conventional fossil plastic resins (Changwichan et al., 2018). With newer bioplastics made from higher generation feedstocks the cost differential can be even greater. Prices for renewable plastics are likely to rise further as demand increases, driven by the need for organizations to fulfil their commitments to reduce or eliminate virgin plastic packaging (Figure 7.10).[19] In addition, higher energy costs threaten the economic viability of recycling facilities. The war in Europe and OPEC production caps have resulted in energy costs escalating to represent 70 per cent of operating expenses in Q3 2022, according to industry

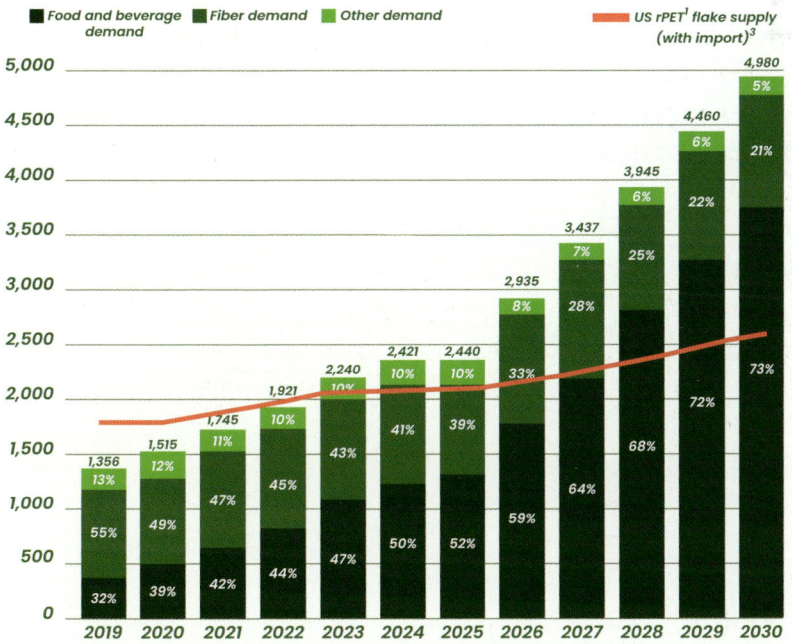

Figure 7.10 Long-term demand for rPET expected to outpace supply in the United States. Adapted from McKinsey & Company (2023).
Notes: 1) Recycled PET, 2) Assuming brand owners pursue at least 50 per cent of public commitments, 3) Assuming importers of food-grade rPET flake increase at the 2017–2020 growth rate.

sources, with some facilities halting operations (Paben, 2022). Those disruptions are also impacting the recycling industry's access to finance. In September 2022, administrators were appointed to Recycling Technologies, a UK-based chemical recycling specialist, after failing to raise sufficient capital to continue operations.

In the case of chairs, the cost of materials often represents a small component of the final retail price. Manufacturers of chairs aimed at the middle and high-end of the market can absorb small changes in the cost of a few kilograms of materials without impacting the retail price. However, in the competitive market for cheap monobloc chairs containing between 3 and 4 kilograms of plastic, the price of materials is very significant. Similarly, producers of many other consumer goods must pass on small increases in input costs to purchasers, potentially impacting demand.

Andreas Maegerlein, Group Leader Creation Center at BASF, also highlighted this as a significant barrier for all new materials, as relatively small quantities are initially produced by material manufacturers at a relatively high cost which must then be passed on to the purchaser. He went on to say that many clients (particularly those looking for packaging solutions) had expressed interest in renewable plastics but quickly changed their minds when they discovered they cost more, highlighting the significance to this barrier in slowing the uptake of these potentially more environmentally friendly materials.

Grcic also told me that those higher costs act as a deterrent for his clients. In the long term, the cost differential for bioplastics is expected to lessen as producers increase production and achieve economies of scale. To succeed, bioplastics (like any new technology) need to prove sufficient demand to encourage large-scale industrial production (R. Kemp, 1994, p. 1024). However, while the International Energy Agency recognizes bio-based feedstock as a potential alternative to oil, they expect bioplastics to remain a niche industry, highlighting the considerable cost gap that bio-based processes need to close to become competitive (International Energy Agency, 2020, p. 57). This forecast is consistent with the market projections presented in Chapter 3, where the bulk of increased demand for renewable plastics is expected to be met by recycled fossil plastics.

3. Lock-in

Lock-in is 'the metaphor to describe actors within a socio-technical regime that gain from perpetuating an existing technology at the expense of a new one, blocking incoming innovations' (Wainstein & Bumpus, 2016, p. 575). Innovations require investment and challenge the status quo, threatening the existence of some actors and imposing significant costs on others. The existing regime, therefore, tends to become institutionalized and resistant to change (Fuenfschilling & Truffer, 2014, p. 772). By prioritizing the maximization of profits for shareholders it is challenging (and often illogical) for organizations to invest in new technologies that threaten to compete with existing capacity (Wainstein & Bumpus, 2016, p. 574).

At the highest level the fossil plastics regime is locked in by decades of investment and experience and it is not surprising that those organizations develop strategies to maintain the status quo. Generous donations are made to political parties and professional lobbyists and industry associations advocate for fossil fuels to influence thought and opinion leaders. Their goal is to safeguard favourable policies, regulations and standards, thereby favouring the existing regime, while hindering emerging challengers.

Manufacturers become locked in by their previous investments in equipment which shapes, directs and influences ongoing research and investment. Many furniture manufacturers outsource some or all of their production processes. Magis outsource to companies in the north of Italy and have built a dependable and reliable local supply chain. During our interview, Ruben Hutschemaekers, the head of marketing and communications, emphasized that this network of suppliers are themselves constrained by their previous investment decisions, making them resistant to change. More broadly, high-volume, specialized production facilities servicing multiple clients from around the world are unlikely to be receptive to requests to accommodate new technology for a single project, especially if the request comes from a relatively small customer. Experimenting with new materials or technologies can then result in additional work to source and connect with new partners and suppliers, a time-consuming process which may cause tension with established the established network of suppliers.

Manufacturing processes are operated by skilled and experienced staff making it expensive to retrain to adapt for new materials or manufacturing technologies. Indeed, staff can become biased toward technologies offering incremental improvement to existing systems rather than those threatening radical change (Elzen, 2004, p. 7). Established operating procedures become locked in by routine and habit (F. W. Geels & Turnheim, 2022, p. 34).

Language barriers and cultural differences can add to the complexity of adapting global supply chains to accommodate new materials. Belgium-based Victor Macadar sourced bioplastic from an Australian supplier and then worked with a Chinese injection-moulding specialist to produce his bioplastic product. His experiences offer insights into some of the complexities an international supply chain can encounter. At first his idea of using bioplastic was met with resistance from the factory, as they preferred to carry on working with conventional plastics, based on familiarity and expertise with those materials. When the factory did agree to develop his design, Macadar encountered further cultural barriers as he explained to me:

> … most moulds are being produced in China … they do not know how to work with this [bioplastics] and this is a social problem. It's not a resistance. It's a lack of capability to work to learn something. It's social, culturally how they teach and [learn]. You give them this material [the bioplastic] and they put in the machine. You give them the parameters and they [enter] in the machine the parameters of traditional plastic … This is a real problem, the most producers are in China, you have to go there and explain them – that is a lot of you lose time. You lose a lot of money because you have to travel that's what happened to me.

Supply chain management and relationship management are separate fields of academic enquiry beyond the scope of this book. However, Macadar's comments highlight the different attitudes and methods of conducting business in different cultural contexts. A representative from the material supplier (Bambacore) agreed with Macadar's observations, adding that in her experience operators in China preferred to learn by doing, using trial and error but this resulted in a significant waste of material. To address the issues encountered by Macadar, Bambacore have since employed a fluent Mandarin speaker in efforts to improve communications between clients and their manufacturers, representing a significant cost increase for the material supplier. Macadar's experience highlights the added complexity of attempting to introduce new work practices into globalized manufacturing networks.

Chiave, from Marcel Wanders' office, summarized the high-level issues confronting designers wishing to work with renewable plastics:

> As much as we designers wants to explore the territory of recycled materials much of the industry (not all) does not because of cost. When you have to start a new production, with new materials it means a new process. It means a new cycle and means investment. So not everyone is very keen for now … The problem is that we need the system around us. We need the system of people and the industry in order to really have a system that is

making recycling work. Our countries, our cities, our environment needs to [be more circular].

Examples of regime pressures locking-in existing investments at all levels of the industry were often cited by designers participating in this research. For instance, Armstrong reported that, despite specifying the set-up of jobs for production, he experienced both domestic and overseas suppliers wanting to deliver projects in their own way, to suit their established manufacturing processes.

Rashid offered an example of how this resistance can manifest when trying to introduce new materials to clients. He told me about a time that he presented to a 'very large company' that produces food goods and snacks to encourage the use of alternative packaging. His suggestion did not receive a welcome reception, as Rashid said: 'a lot of these companies don't want to hear that from me. It's kind of like this isn't really your business or this is our supplier'.

Rene Linssen, an industrial designer at Formswell, is eager to use bioplastics but has heard reports from others who had taken five years to develop products in-house. He viewed the material as inaccessible to him, citing the associated development costs and claiming that those who had successfully experimented with bioplastics wanted to protect their investment and not share what they had learnt. Linssen makes a particularly important point with that statement as the MLP model is dependent on continued experimentation with an emerging technology, and an unwillingness to share their findings impedes future experimentation.

Designers are also locked into a model of industrial design focused on the artefact. Clients commission designers to develop specific products, a brief might specify a niche market, such as education, but patrons do not usually instruct (or pay for) designers to interrogate the precise requirements of educational establishments and develop alternative solutions to satisfy those needs. While designers continue to be engaged on short-term contacts as 'service providers' to create artefacts they have limited opportunity to go beyond their brief and investigate radical alternatives such as product-service systems (PSS) which might deliver better environmental outcomes.

a. Industry consolidation

Industry consolidation can further exacerbate lock-in effects, as larger organizations require more investment to modify or adapt their existing infrastructure.

Some plastic furniture manufacturers remain independently privately owned (Vitra, Kartell, Vondom and Magis among them), but they are exceptions. The constant pursuit of growth and improved profits demands ever larger economies of scale which drives mergers and acquisitions across manufacturing. In the furniture market the Italian conglomerate Investindustrial acquired furniture producer B&B Italia and lighting manufacturers Louis Poulsen and Flos creating a new company, Design Holding. The American office furniture specialist Haworth acquired Poltrona Frau (who owned Cappellini) and Cassina, creating one of the world's largest companies dedicated to furniture production, since adding Interni, the Luxury Living Group (which produces furniture lines for Versace, Dolce & Gabbana, Bentley and Bugatti) and Zanotta to their portfolio (Hahn, 2023; Ross, 2020). In 2021, Herman Miller acquired Knoll and rebranded as MillerKnoll, since adding Hay and Muuto (among others) to the stable. This trend is not unique to the furniture market – constant pressure to grow drives the acquisition frenzy. The larger a company gets the more constrained by convention it becomes, as the costs associated with change rapidly escalate.

4. Summary

The barriers confronting both manufacturers and designers seeking to work with renewable plastics formed the focus of this chapter. The emphasis here is on factors that influence the choice of materials, where the manufacturer typically holds the balance of power. Despite not usually having the final decision the designer often does play a crucial role in encouraging experimentation with renewable plastics. Designers can be agents for change.

The first major obstacle to the adoption of renewable plastics is the difficulty in obtaining up-to-date and independently verifiable information about new materials and where to source them. As many of these materials are new, they are only available (often in limited quantities) from specialist suppliers that can be hard to find. The ever-increasing variety of bioplastic and hybridized products available is adding yet more complexity to an already-confused market. Designers interviewed for this study highlighted a lack of relevant information sources tailored to their specific needs. The lack of transparency from material manufacturers, combined with the abundance of misinformation in the market, some of which has been deliberately circulated by the incumbent regime, complicates the search for impartial information.

Bioplastics are significantly more expensive than traditional plastics and while that price differential is expected to erode as production expands, the premium is unlikely to be eliminated entirely. Many bioplastics are designed to biodegrade, in response to the needs of the packaging industry. Bioplastics suitable for durable consumer goods destined for prolonged use remain comparatively scarce. Material manufacturers are faced with many challenges in their efforts to grow demand for bioplastics tailored for use in consumer products and therefore use of these materials is expected to remain rare, until costs are reduced, and their performance improves.

Those interested in working with renewable plastics are also likely to encounter resistance from suppliers and others who are effectively locked in to virgin plastics by their previous investments in both infrastructure and skills. Designers and manufacturers who have completed projects using renewable plastics warn that significant investments of both time and money is required to gain the necessary skills and expertise. This need for investment in time, combined with the higher cost of renewable plastics, especially bioplastics, has implications for retail prices, threatening to constrain demand for products based on these materials.

Despite the formidable list of barriers confronting those seeking to experiment with renewable plastics many designers and manufacturers have already developed products that celebrate the use of these more environmentally friendly materials. The following final chapter identifies strategies that can be adopted by actors interested in driving the widespread adoption of renewable plastics.

8 Transition, how?

'To be' we have to be another way.
Tony Fry (Fry, 2009, p. 22)

From the discussion of strategies to reduce the GHG impacts of plastics discussed in Chapter 3 it was shown that the only viable short-term solution is to increase demand for renewable plastics, thereby reducing demand for virgin plastic. But how can environmentally successful innovations using these materials at the product level be scaled-out and scaled-up? How can valuable lessons from the successful niche experiments identified in Chapter 4 and discussed in Chapter 5 be applied more broadly if individual products fall foul of vagrancies of fads and fashions? How can industry participants be persuaded that the barriers to using renewable plastics discussed in the previous chapter can be overcome and to adapt or even abandon manufacturing processes that have been refined through decades of investment?

MLP identifies strategies that have successfully supported transitions in other industries. In this chapter these strategies are examined for their relevance in expanding the use of renewable plastics in consumer goods. Transitions typically take decades to occur, but this is simply time that we do not have, if we are to avoid the projected devastating environmental consequences accompanying the fossil regime's current growth targets. Identifying and applying strategies and tactics to accelerate the transition to more sustainable renewable plastics is vital.

Despite compelling pressure to continue business as usual, the continued use of virgin plastics will face escalating scrutiny, as the health and environmental impacts of the material become more increasingly evident. Renewable plastics are poised to gain greater favour with manufacturers as demand from purchasers increases. As the era of post-extractivism approaches, the petrochemical industry will have no alternative but to find another way to be.

1. Strategies to accelerate transition

Previous Multi-Level Perspective (MLP) studies have identified several strategies that are common to challenger technologies that have successfully replaced incumbents. Examination of case studies of the designs that achieved the highest ratings using the Environmentally Responsible Product Rating (ERPR) tool (discussed in Chapter 4) confirm that many of these same strategies are relevant to those interested in promoting the use of renewable plastics. Strategies are categorized under five headings: hybridization, targeting high-growth market segments, multi-technological impacts, networks and timing (Figure 8.1).

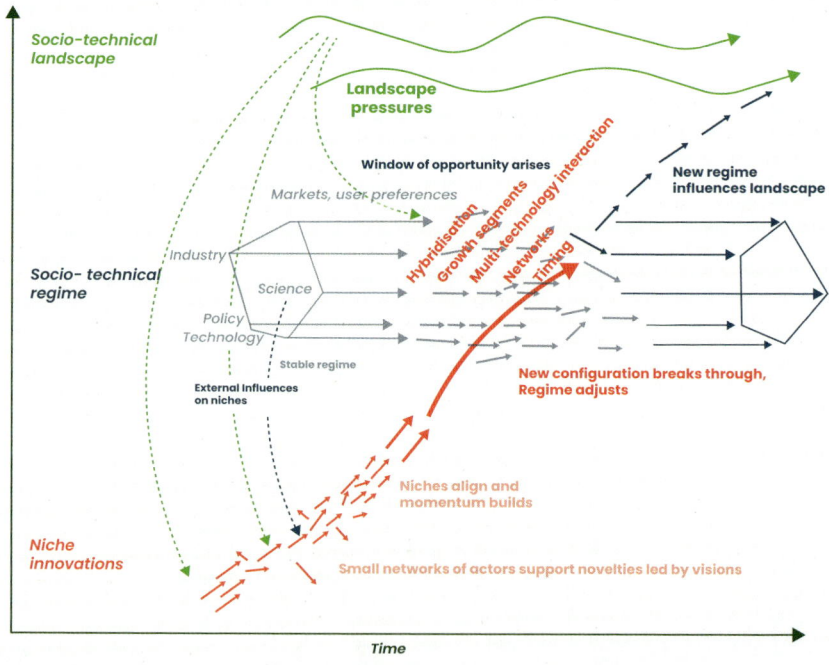

Figure 8.1 Regime breakthrough is achieved, and the landscape adapts to absorb the new technology. Adapted from Schot & Geels (2008).

a. Hybridization

New processes or solutions that can easily integrate into existing mainstream practices with minimal disruption are more likely to be taken up (Smith, 2007, pp. 444–446). Hybridized plastics can simplify the process of locating a material with the mechanical specifications required for any given project.

Hybridization has already been tested and proven as a very successful strategy for fossil plastics to infiltrate existing regimes. Natural fibres such as wool, cotton and silk have been largely displaced by synthetic fibres, thanks to a growing acceptance of less-expensive hybridized offerings such as poly-cotton, increasingly available since the 1930s. Synthetic fibres now account for two-thirds of all fibres produced globally demonstrating the success of hybridization in this market (European Environment Agency, 2021, p. 16).

Most recycled plastics available on the market today are essentially hybridized, with virgin polymers added in efforts compensate for any loss in quality caused by recycling processes. This process also minimizes variation in properties found across different grades of material classified as the same type and helps maintain a consistent melt flow index (Sin & Tueen, 2022, p. 35). Hybridization enables recycled plastics to display almost identical physical characteristics when compared with their virgin counterparts. Manufacturers and designers can use these materials with relatively minimal adjustments to their existing processes. Investments in both time and money to retrain staff and re-tool or recalibrate production facilities are minimized. It is unsurprising, then, that major manufacturers such as IKEA and Vitra have made their first moves into renewable plastics by using recycled materials. IKEA introduced the *Odger* in 2017 made from recycled post-industrial PP mixed with wood, while Vitra released a version of Barber Osgerby's *Tip Ton Chair* made from recycled post-industrial PP in 2020 (Figure 8.2). In January 2024, Vitra announced plans to use recycled household waste PP to replace virgin plastics across its range, commencing with the Eames Shell Chairs which, they claim, can be recycled at end-of-life.

Rather than offer hybridized or fully recycled plastics as separate products, material suppliers are promoting a 'mass balanced' solution for clients. To fulfil orders for these materials, an equivalent quantity of plastic (adjusted for yields and conversion factors) is purchased from the waste stream. The discarded plastic is chemically recycled (treated by pyrolysis) with the resultant oil mixed with traditional fossil inputs and polymerized to produce plastic (BASF, 2022).

Figure 8.2 *Tip Ton RE*, Barber Osgerby for Vitra, 2020. Made from recycled PP, PE and steel. Courtesy Vitra.

Alternatively, bio-derived oils can be mixed to displace the use of fossil fuels. The client can then (using the free allocation method) claim the renewable content of the material based on the proportion of feedstock which has been allocated to their product. An independent auditor's certificate is supplied to support those claims.

Mass balancing is attractive to clients as the substantial costs of adapting existing manufacturing processes to match the mechanical characteristic of a different material are avoided. Small quantities of recycled or renewable organic material are mixed with a large quantity of virgin fossil material meaning that the output is indistinguishable from the equivalent 100 per cent virgin plastic. Clients can potentially claim their product as being made from up to 100 per cent recycled feedstock when, in fact, products actually contain a minimal quantity of recycled ingredients. However, allowing products made from this process to be labelled as made from renewable plastics is contestable.

The mass-balance approach is relatively complicated to communicate to clients, let alone end-users. Maegerlein from BASF admitted as much to me in our interview. Axel Barrett, the editor and publisher of *Bioplastics News*, goes further calling it a 'Rubiks Cube Puzzle'. He claims to have 'no clue what it means. They inject biobased feedstock on one side of the galaxy, and they eject fossil-based stuff on the other side of the galaxy but it's branded (bio) "mass balanced"' (Barrett, 2022a). Even the petrochemical industry is calling for standards to be developed to implement mass balance in LCA methodology (Together for Sustainability, 2022, p. 79). Certainly, the complexity of the process could result in marketing messages that simplify or overlook details of the process thereby misleading purchasers, particularly those interested in selecting products based on their environmental credentials. While it can be argued that the actions of the product manufacturer have resulted in an equivalent decrease in demand for virgin plastics, the fact remains that only a small proportion of the material used to make the product is actually recycled feedstock or made from renewable organic sources.[1] There is a risk that purchasers will unknowingly select products containing significant quantities of virgin plastic sourced from fossil fuels. Current quantities of mass-balanced feedstock are negligible. In 2020, BASF only processed 1,000 tonnes of recycled materials, but aims to increase this to 250,000 tonnes by 2025.[2]

Bioplastics are also often hybridized with virgin plastics to improve compatibility with existing manufacturing processes and to compensate for any deficiencies in mechanical properties. Recycled plastics are usually only mixed with virgin material of the same type meaning they can, at least theoretically, be recycled at end-of-life. Mixing bioplastics with fossil plastics to create bio-based materials results in poor end-of-life outcomes, as bio-based plastics cannot be recycled and do not decompose. Despite these limitations, production of bio-based materials reduces demand for at least some virgin plastic, as renewable organic material has been used in its place. If the material has been used to manufacture a consumer product designed to last for many years, the end-of-life impacts are postponed, adding to the short-term benefit of reducing demand for fossil fuels.

Hybridization is also often appealing to end-users as well. By mixing familiar technologies with new, the market is given a stepping stone, an opportunity to build confidence in the new, as the resultant products are not radically dissimilar to the familiar. However, adoption is more likely to succeed where unique and immediate benefits are evident to potential purchasers. Blending cotton with polyester produces a breathable, tear-resistant fabric that is less expensive than natural fibres and easier to iron. Automobile manufacturers are adding features such as lane control or assisted parking to their vehicles demonstrating the convenience of automated driving while building market confidence in the technology. Designers may struggle to identify any specific consumer benefits associated with hybridized plastics. However, as we have already seen, some designers have taken the opportunity to differentiate products by showcasing the unique aesthetic qualities of these new materials (Chapter 5). It remains to be seen if these new aesthetic and sensory experiences offer sufficient consumer benefit to drive the update of renewable plastics in consumer goods.

b. Targeting high-growth market segments

MLP studies have shown that new technologies can transition to the mainstream by identifying and servicing high-growth niche market segments. As the niche market develops, the new technology is propelled with it. Geels and Schot observed that steamships benefited from strong growth in the Atlantic passenger market, with historical events such as the potato famine in Ireland and the American gold rush driving demand. The need for faster, more reliable cross Atlantic transfers for people created a new niche market ideally suited to steamships. The speed and reliability improvements provided by the new technology justified a higher price point (F. W. Geels & Schot, 2007, pp. 410–411).

The early adopters of products made from renewable plastics are most likely to be found among environmentally conscious consumers. Although they might be identified by surveying the market for consumers' beliefs and attitudes, environmentally conscious buyers are difficult to reach as they cannot be accurately targeted using traditional demographic segmentation when planning marketing activities. Three extensive reviews of relevant academic studies investigated the link between demographics and potential to purchase more sustainable products and found little evidence of a correlation (Diamantopoulos et al., 2003; Horani, 2020; Joshi & Rahman, 2015).

All three of these literature reviews, together with the results of original research conducted for this book, highlight the challenge of targeting purchasers interested in environmentally conscious design. Many consumers often hold negative associations with sustainable products, viewing them as being of lower quality, less aesthetically pleasing and more expensive (White et al., 2019, n.p.). Marketing the sustainable credentials of products to this audience wastes resources. Effectively targeting environmentally conscious purchasers with tailored messaging specific to a particular product category precisely at the time a purchase is being considered is extremely challenging. Many consumer goods might only be acquired once every decade or even less often.

Despite these challenges, some designers and manufacturers have already succeeded in targeting discrete market segments with interest in sustainability. In the case of chairs, hotels, conference centres and restaurants were identified as early adopters of renewable carbon creations. Pedersen reported that, within the last two or three years, he had noticed increased interest in environmentally friendly products from restaurants and hotels in Europe, whose commitment to lessening their environmental impacts extended beyond asking guests to reuse their towels. The first big order for *Falk* (made from recycled household plastic waste) came from a hotel who selected the chair as 'it was the best on the market' in terms of design, comfort, price and especially because of its environmental profile. Purchases by hotels and restaurants are visible signifiers reinforcing the organizations ethical commitment to sustainability and providing them with a point of differentiation to compete in a crowded market. Designers and manufacturers benefit from sales directly and indirectly, from having their work exposed in public spaces where the chairs become part of the cultural experience enjoyed by large numbers of potential customers. Those hotels, restaurants and conference centres that choose sustainable alternatives help demonstrate that luxury and sustainability are not mutually exclusive. By showcasing sustainable furniture, they act as opinion leaders, contributing toward social norming of renewable carbon products and encouraging wider adoption.

Pedersen also highlighted schools as a potential target market, noting that governments are increasingly concerned with integrating sustainability in their purchasing policies. Governments are taking action to mandate the use of recycled materials across all areas of activity. As the focus on sustainable procurement becomes commonplace in more jurisdictions around the world governments have the potential to generate significant demand by implementing supportive guidelines and regulations. In turn, this creates opportunities for designers and manufacturers engaged with developing more sustainable solutions across many product categories.

The chair market continues to fragment, with chairs adapted for increasingly specific niche markets. In the office, for example, typists and executive chairs have been joined by drafting chairs, kneeling chairs, sit-stand chairs, swivel chairs, mesh chairs, balance ball chairs, saddle chairs, active sitting chairs, conference chairs, reception chairs, tablet armchairs, folding chairs, chairs for outdoor formal and casual settings, and many more. With chairs specifically designed to meet the needs of these users it appears optimistic to launch another general-purpose chair. However, most of the designs discussed in this book were apparently created without reference to a specific target market. Designs are more likely to be promoted for their versatility – suitable for both indoor and outdoor use across both the residential and business markets, for example.

Several of the chairs I analysed were designed to allow prospective purchasers to select from a range of bases and shells to tailor the chair to meet their specific needs. Oki Sato adopted that approach for his *NO2 Recycle* design for Fritz Hansen; also offering unupholstered or upholstered options in efforts to appeal to niche markets within both the commercial and residential sectors (Figure 8.3). Offering a wide range of configurations theoretically improves the appeal of the

Figure 8.3 *NO2 Recycle*, Oki Sato for Fritz Hansen, 2019. Showing a selection of the colours and variants offered to tailor the *NO2 Recycle* to the needs of various niche segments across the business and residential markets. Courtesy Fritz Hansen.

chair (and contributed to Sato's design achieving a high score in my ERPR analysis). However, that approach is also subject to the risk inherent in any compromise – failure to fully satisfy the needs of some, or even all, potential purchasers. While acknowledging these risks I assumed they are outweighed by the benefits of offering greater variety when scoring these designs using the ERPR tool.

Surprisingly, only three of the designs included in my analysis appear to have been developed with a specific target audience in mind. *Charlie Chair* was designed as a chair for children, the *Tip Ton* (precursor to the RE version) was designed for the education market, while the *Bell Chair* is specifically aimed at a younger, design conscious audience (twenty to thirty-five-year-olds). Both Konstantin Grcic and Magis told me that need for the *Bell Chair* needed to be stylish and inexpensive to appeal to that market was explicitly stated in the brief. The articulation of a target market, together with the specification of a target price point, provided a clear direction for the design solution. Backcasting from the retail price determined the allowable production time and the amount of material that could be used. Matching the design to the needs of a specific market is a proven strategy for commercial success. Although launched halfway through the year, Hutschemaekers claimed the *Bell Chair* to be among the top three best-selling designs for Magis in 2020.

Adopting a more targeted approach and meeting the precise needs of a specific group of consumers, embracing consumption within the domain of design, could offer the opportunity to drive the take up of renewable carbon-based chairs. The client is responsible for specifying the target market for their product, but it is in the best interest of the designer to ensure the brief clearly articulates the needs of a specific market. While the commercial success of any design will be ultimately determined by the market, tailoring designs to satisfy the needs of a specific niche market follows a marketing strategy proven to deliver more predictable results that will accelerate transitions.

While determining a product price is beyond the remit of most designers, their decisions affect costs, which are reflected in retail prices. Renewable plastics (and bioplastics in particular) often cost more than traditional plastics and are likely to for the foreseeable future. Quitllet outlined the issues driving costs, resulting in increased prices for his designs featuring recycled plastics:

> [Working] with a recycled material is a most difficult process. There is much more time to do it and to have it right because there is more mistake pieces that you need to recycle again. And with a higher price and with a less beautiful finish because it's not completely regular, it's nice because

I like to know there's this story inside that chair. Every chair is different, because there's a different part of the story inside. Not like the others, which are all the same but you need to put this point of view and say okay, if I want this beautiful chair you have to pay.

In this statement Quitllet acknowledged the role that the aesthetic appeal of the recyclates together with the provenance of the material can play in justifying the price premium to end-users. Higher retail prices can be expected to stem demand. However, high prices do not necessarily have a negative impact on the uptake of new technology. Higher retail prices imply higher margins for retailers and distributors, incentivizing them to carry and promote stock. Successful higher priced niche experiments will attract interest from competitors in search of higher profits.

For those high-priced products to sell, their environmental credentials must be accepted, valued, preferred even, by consumers. But are buyers prepared to pay more for sustainable designs? Several studies across different jurisdictions consistently identify a sizeable segment of the market prepared to pay more for sustainable designs (Nova Institute, 2020a; White et al., 2019). Encouragingly, a study undertaken in the Netherlands found consumers particularly willing to pay a price premium for products containing recycled ocean plastics (Magnier et al., 2019). Although the results from these studies are promising, they should be treated with caution. Any survey of consumer attitudes only reflects the opinions of participants drawn from a (usually limited) specific geographic location at a particular time and place. Results from a 2019 survey of Dutch consumers cannot be extrapolated to other markets or other time frames. Low-income markets are less likely able to pay a premium.

The designers who spoke with me expressed a spectrum of opinions when asked for their view on their ability to charge a premium for sustainable products. Some, including Gabriele Chiave, found consumers 'willing to spend a bit more for something less disposable, more sustainable'. He then emphasized the need to provide provenance around the material to justify the additional cost. Victor Macadar agreed, saying his customers valued the biodegradability offered by his products and are prepared to pay a premium. Sarah Gibson reported a different experience reporting that their customers are not prepared to pay more for sustainability, offering a specific example:

I'm thinking about the *Ribs Bench*, which was designed by Stefan Lie. When we took it onboard it used to be made out of MDF strips, laminated together. We worked to find a solution without laminates and resin, and I don't think anyone cares about them. We like it and I don't think we [would] want to sell it if it wasn't made like that but I don't think that's why people pay premium – they pay for the design and shape and all that.

Grcic reported that attitudes to sustainability had shifted in recent years and is now considered as contributing to the quality of a product for which 'people are very happy to pay a premium'.

Pedersen differentiated between the residential and business markets reporting that he found companies are prepared to pay more for recycled materials, but the mass residential market is not. During a site inspection of an IKEA store, Kate Ringvall, Sustainability Business Partner, expressed her frustration at the reluctance of the display team to use point-of-sale material emphasizing the sustainability claims of some products. The display team, she said, believe there to be little consumer interest in such messaging and therefore focus their attention elsewhere. Similarly, information about the sustainability credentials of the material used to create the *Odger* is not available on the company's website (IKEA Australia, n.d.).[3] This absence is particularly disappointing given IKEA's very public focus on 'sourcing and producing renewable and recycled materials with a positive environmental impact' (IKEA, 2020, p. 15).

While prices can be compared in absolute terms, the real competitiveness of any product can only be evaluated within the context of its specific market conditions. For example, the *Odger* was available for AU$99 at IKEA in Australia during 2021 but the price has since increased to AU$179 and then reduced to AU$149.[4] While the price remains competitive when compared with the other chairs I have analysed, the *Odger* is expensive in comparison to other similar chairs available from IKEA (with prices starting at around AU$39 and a wide variety of models available at prices up to AU$75). If seeking to purchase six or eight dining chairs, the *Odger* effectively commands a significant premium for its sustainable credentials. While the environmental benefits of the design remain unpublicized, prospective purchasers will remain unaware of the reasons for this price difference and be left with little justification to pay the premium.

Given the complexities around pricing and a lack of data on sales volumes it is difficult to produce quantifiable evidence to evaluate the premium purchasers are prepared to pay for any sustainable product. What is clear is that for renewable carbon products to gain significant market share in any product category end-users must be encouraged to value and be prepared to pay for renewable plastics. The true cost of the defuturing implicit in products made from virgin plastics must be made explicit. In a press interview industrial designer Stefan Diez claimed that it is the responsible consumer who will finance the development of more environmentally friendly designs, to the benefit of everyone in the long run (Moldenhauer, 2021). Mass-market opinions change over time and add to demand for radically new technologies as the benefits of these disruptive innovations become more widely understood (Penna & Geels, 2015, p. 1031). This emphasizes the responsibility of all those involved with the development of more sustainable plastic products, from design to distribution, to encourage prospective purchasers to consider the true costs of fossil plastics to both the planet and its people.

i. Marketing

The development of marketing messaging is most likely beyond the direct influence of most designers; nevertheless, emphasizing the environmental benefits of their creations can provide valuable input to the development of the sales narrative. Designers can take the lead and inspire those involved with the promotion of products. In my interview with Grcic, he explained how he worked with Magis to develop a separate micro-website specifically to explain the story of the *Bell Chair*. Based on examples from the food and fashion industries, the aim was to fully explain the source of material and the efforts made to optimize production to improve the environmental credentials of the chair. Grcic worked closely with the marketing team at Magis to finesse both the content and tone of messaging and to ensure copy did not include exaggerated claims. Opportunities for freelance designers to become this deeply involved with marketing are rare, as projects usually end when designs enter production and economic necessity demands the designer moves on to the next project. However, among the chairs included in my study Grcic and Magis set the standard for what can be achieved from a close collaboration between designer and manufacturer's marketing team.

In efforts to imbue a product with personality and to foster an emotional connection with end-users, marketing and advertising teams may endeavour to instil a perception of historical lineage or a rich design heritage. Story telling around provenance is a strategy often featured in the sales narrative for products made from traditional materials. For example, 'handmade from sustainably sourced oak from the USA'. Alternatively, the craft skills used during construction might be emphasized: 'handcrafted using sheep wool and traditional techniques'.

During our interview Australian designer Trent Jansen explained the commercial pressures that he has experienced to provide detailed information about the materials he uses:

> My gallerist constantly asks me about the provenance of the materials that I'm working with. 'Where's this from? Who made it?' She wants selling points … That's something that sort of adds to her narrative and something that her audiences are interested in. I don't think unless the project was about innovation in plastic or using plastic in a in some kind of really interesting [she would be interested]. She never asked me, 'Where did the plastic feet come from? What's the provenance of plastic?'

Jansen's comments highlight the important role for designers in communicating the provenance of material to other actors involved with a project. Where fossil plastics are used as the primary material, those considerations are usually considered irrelevant. Renewable plastics can differentiate themselves from fossil plastics by celebrating their history.

When we encounter antiques, they contain the essence of their former lives and experiences; designer and academic Nick Grant argues that recycled materials also develop and gather importance through their extended lives (Grant, 2017, p. 225). He goes on to suggest that the recycling process can form part 'of the articulated language and sophisticated narration that becomes embodied in the final object' (Grant, 2017, p. 227). He argues that materials themselves have agency and offer the ability to mediate the message of a more sustainable society. Use of recycled material can act as a reminder of the collective value and our responsibility implicit in managing scarce resources. Magis attempt to leverage those associations by promoting the fact that the material for the *Bell Chair* is derived from waste generated by their own furniture production, supplemented by waste material sourced from the local automotive industry. These localized sources then form part of a detailed narrative around the sustainable properties of the *Bell Chair*.

The emotional response to the ocean plastic crisis is being strategically harnessed to construct powerful marketing narratives in deliberate attempts to position products as viable solutions to address the challenges posed by ocean plastics. For example, the Vondom website promotes the *Revolution Collection* (Figure 8.4) claiming:

> The soft white tone is the result of our work closely with companies specialised in the collection and recovery of plastics from the Balearic Islands and Ibiza. Subsequently and together with transforming companies, we managed to convert these resources into quality new material, recycled plastic with a natural tone.
>
> (Vondom, 2021)

The accuracy of any story told around provenance depends on who wrote it and how thoroughly any third-party claims have been investigated. Examples where chairs are marketed as being made from 'ocean plastics' provide illustrations of many dubious claims being made by some manufacturers (unlike Vondom). Unsubstantiated or misleading claims can destroy trust when they are discovered, negatively affecting attitudes to sustainable eco-design more generally. Indeed, several studies conclude that the use of eco-labels is already of limited value, as trust has been destroyed following the introduction of a plethora of systems designed to indicate the environmental profile of products.[5] While many of these schemes were developed by organizations with a bona fide interest in promoting sustainable purchasing, others merely represent greenwashing attempts by manufacturers. The quantity and complexity of these schemes (and the dubious claims

Figure 8.4 *Love*, Eugeni Quitllet for Vondom, 2018. Part of Vondom's *Revolutions* series made using 'recycled fishing nets from the Mediterranean Sea'. Courtesy Vondom.

made by some users) has left purchasers confused and sceptical to environmental messaging.

More positively, provenance narratives can help foster a bond between the user and the product. With storytelling designed to evoke the ghost of the material's previous life, purchasers are encouraged to take pride from their environmentally conscientious choices in acquiring these products. The history of the material has added meaning to the product. A 'circle of virtue' can be enhanced when a consumer can configure a sense of the material being returned to them (Grant, 2017, p. 226). Designers can author authenticity by embodying recycling activities within the physical attributes of their products, thereby engendering a more ethically principled interpretation of the materials used. Dutch designer Dirk Vander Kooij, who repurposes the internal components of refrigerators to craft his *Chubby Chair* (Figure 8.5), agreed with this, claiming sustainability considerations are not his primary motivation, but rather, he uses the recycled plastic due to its distinctive historical narrative, which he regards as the most compelling aspect of the material (Day, 2011).

Working with recycled plastics creates the opportunity to focus attention on the downstream impacts of the material and promote the circular economy. On the other hand, bioplastics offer the potential to contrast the upstream impacts of fossil plastic production. Design academic Ann Thorpe observes that

Figure 8.5 *Chubby Chair*, Dirk Vander Kooij, 2012. Courtesy Studio Dirk Vander Kooij.

few eco-design approaches are afforded the opportunity to link consumers to the upstream social and environmental consequences of making products, observing: 'many designers are as distant as consumers from these upstream effects' (Thorpe, 2010, p. 7). By examining upstream impacts designers and marketers create the opportunity to differentiate bioplastic products from both virgin and recycled plastics.

Designer Nikolaj Carlsen offers extensive information on the natural source of the seaweed composite used for his *Coastal Furniture* chair (Figure 8.6); he also emphasizes the historical use of the material to demonstrate its proven durability. Carlsen has created a story illustrating how the material has been used to thatch

Figure 8.6 Provenance of material used for *Coastal Furniture* as explained by the designer: 'The seaweed roofs on the island Læsø (Denmark) has been the key inspiration source. The shell for the lounge chair is made of one hundred per cent biodegradable seaweed composite … The material consists of eelgrass and carrageenan. Eelgrass grows naturally near the coastline only to be collected when washed ashore. Carrageenan is extracted from certain red algae' (Carlsen, 2019). Courtesy Nikolaj Thrane Carlsen.

roofs on a Danish island for centuries. He then goes on to explain the environmental benefits of the material to instil an emotional connection with the chair (Carlsen, 2019).

Carlsen's emotionally charged approach contrasts with the basic facts often used by the manufacturers of the other bioplastic chairs I studied. For instance, Braskem highlight that the bioplastic they supplied to Tramontina for use in the *Jet* series of chairs result in the capture of 3 tonnes of CO_2 per metric ton produced (Braskem, 2018). A Lot of Brasil emphasize the regenerative properties of the trees used to make Rashid's *Siamese Chair* (Rodriguez, 2016). There is potential for designers and manufacturers to develop more engaging stories explaining the natural ingredients sourced to create the bioplastics they are working with. Bioplastics developed from second-generation (or above) sources, particularly when they have been processed using sustainable energy, cause far lower environmental impacts when compared with fossil plastics, but this needs to be explained to end-users. Emotionally engaging narratives can be constructed to highlight supply chain differences and the cascading environmental benefits of purchasing products made from these materials.

Investigations into the provenance of virgin plastics can also be used to reinforce the selection of renewable plastics by emphasizing the omnipresent unmaking concealed within fossil plastics. On arrival at an academic conference, attendees were offered a carton of ice-cream (Marriott & Minio-Paluello, 2013). After the ice-cream had been consumed, the presenters began to explain the destructive 'pre-life' of the plastic used to make the container, tracing:

> The passage of that material from oil-bearing rocks, through drilling rigs, pipelines, terminals, depots, refineries, factories, distribution centres and shops, to homes … Examin[ing] the impacts – both ecological and social – of that passage.
>
> (Marriott & Minio-Paluello, 2013, p. 172)

While acknowledging that it is impossible to know the precise source of the materials used to make the carton the most likely scenario was presented. Oil extracted in Azerbaijan is transported by pipeline and sea to a processing plant in Germany where a small portion of it is turned into plastic before making its way to the ice-cream factory in the UK. In detailing the journey, we are confronted with some alarming details about the destructive activities and questionable employment practices of those involved with the industry. The authors conclude by highlighting that the true costs of disruption and violence caused by the extraction of fossil fuels remains largely invisible to the end-user. This is especially true of plastics, where users often remain ignorant of the resources consumed in their production.

With gas from shale activities in the United States becoming an ever-greater source for plastics (even for Europe), it is incumbent on those using these materials (and products subsequently made from those materials) to familiarize themselves with their 'brutish origins' and the consequences of the environmental racism caused by their creation. However, the different feedstocks used to create fossil-based plastics are often ignored by those promoting plastic products. The chemical composition of PP (or any specific plastic for that matter) remains unchanged regardless of the feedstock source and is usually regarded as homogeneous. Nonetheless, the environmental impacts associated with their creation vary depending on the source of the feedstock. Investigation of the provenance of fossil plastics remains a rich and unexplored territory to discourage the use of the material. For instance, China still relies heavily on coal to produce plastic (Chapter 4). When used as a feedstock for plastic, especially when also used to generate the energy required to process it, coal creates the largest CO_2 emissions compared to alternative fossil feedstocks (Bennett, 2012a, p. 331). This environmental impact remains a hidden cost of often the lowest-priced plastics, and warrants consideration by designers, manufacturers and purchasers alike.

Publicizing the negative upstream impacts of fossil plastics can highlight the conceptually and geographically distant costs of irresponsible consumption and undermine the incumbent regime. By revealing the source of the materials they work with, and telling the story of their creation and unmaking, designers and manufacturers are uniquely placed to promote the adoption of renewable plastics. The main challenge here is to communicate this information in a simple, trustworthy and effective format.[6] Successfully communicating direct consumer effectiveness has been demonstrated to positively correlate with green purchase intentions (Joshi & Rahman, 2015, p. 137).

Individual designers or manufacturers might consider the role of educating consumers about the upstream impacts caused by the creation of fossil plastics as beyond their remit. However, many manufacturers already use Forest Management Council® certification to reassure consumer of the origins of their timbers. A similar opportunity exists when renewable plastics are used with the Nova Institute already making one suggestion for an equivalent symbol of endorsement (Figure 8.7).

Figure 8.7 Forest Stewardship Council® logo (left) indicates timber used in a product has been sustainably sourced. Renewable carbon logo developed by the Nova Institute (right) has potential to be used by industry to signal the use of plastics that meet sustainable criteria. Courtesy FSC® and Nova Institute.

c. Multi-technology impacts

Renewable plastics have already branched into other industrial sectors having been primarily developed to meet packaging demands. The automobile industry has been an early adopter of those materials, and manufacturing processes have already been adapted to accommodate them (Jeong & Ko, 2020, p. 136). Bioplastics have been met with enthusiasm by that industry as the use of eco-friendly materials reduce CO_2 emissions, helping manufacturers to meet sustainability commitments.

Two of the chairs achieving the highest scores in this study, the *Hemp Chair* (Figure 8.8) and *Nobody Chair* (Figure 8.9), are moulded in one-piece using the same thermo pressing process used to form parts for cars, such as removable boot shelves. Compression moulding consumes less energy than injection moulding. The relatively low pressure required by this technique delivers additional benefits: the cost of expensive steel moulds is avoided, enabling smaller production runs to be economically viable. As more bioplastics enter the market this energy saving technology has potential to become increasingly common, particularly among small and mid-range furniture manufacturers.[7] Both the *Hemp Chair* and *Nobody Chair* are stackable, demonstrating that compression moulding also opens new horizons for designers, in these instances stacking chairs have been created without resorting to the traditional solution of cutting holes in the back of the shell.

Figure 8.8 *Hemp Chair*, Werner Aisslinger for Moroso, 2011. Illustrating the choice of colours available for this stacking chair. Courtesy Alessandro Paderni for Moroso.

Figure 8.9 *Nobody*, Boris Berlin & Poul Christiansen for Hay, 2007. Compression moulded using felt made from recycled PET. Courtesy Boris Berlin.

Other designers have combined multiple innovative technologies in attempts to gain synergy, exceeding the benefits that can be achieved by using them individually. Starck used AI to create a chair for Kartell (Figure 5.21). Dirk Vander Kooij purchased an industrial robotic arm, retired from the automobile industry, and added an extruder to develop his *Chubby Chair* (Figure 8.5) This interaction of 3D printing and robotics is similar to the approach adopted by Marcel Garcia for his *Voxel Chair v1.0*. Grcic pointed out to me that the automobile, aeronautical and medical industries already use 3D printing to produce complex components for products and that this approach could be the future of the technology in the furniture industry. Alternatively, designers might utilize 3D printing for some components and combine with different manufacturing technologies to create the remainder of a product. Both additive manufacturing and compression moulding offer the advantage that they are both supportive of continuous improvement efforts, as modifications and enhancements can be made to designs without incurring the prohibitive costs associated with replacing steel dies for injection-moulded projects.

Experimenting with digital technologies allows the designer (or student) to become more deeply involved in the production process and has the potential to highlight sustainability issues. Stefan Diez attributes his own interest in sustainable design to the development of rapid prototyping. Completing his training at a time when 3D printers and computer-controlled manufacturing machines were emerging allowed him to become engaged in the product-development process. This experience fostered a profound understanding and appreciation of the resources required for a project (Moldenhauer, 2021).

New manufacturing technologies prompt the re-evaluation of long-standing solutions to common design problems. For example, rectilinear ribs are often used to add strength to chairs. Making straight cuts in a mould with a milling machine are an easy (and cost-effective) solution when working with injection moulding. With other technologies, where the upfront investment in an expensive high-pressure mould is not required, designers are afforded the liberty to engage in exploratory endeavours and experiment with less-intrusive solutions. Structures that occur naturally in our environment provide a variety of models to inspire solution to these challenges. The designers working with these emerging manufacturing technologies are only just beginning to explore their inherent potential.

d. Networks

Any transition is beyond the unilateral control of a single actor, as niche participants lack the agency required to effect regime change. For transitions to occur, individual participants must construct coalitions of actors, thereby aggregating the requisite skills and resources necessary to instigate meaningful change. Knowledge must be shared across formally and informally to promote intersectoral cooperation across niche market segments. Coalitions of actors can be nurtured by developing a detailed understanding of the motivations, interests and expectations of each actor, allowing a pathway for co-evolution to develop.[8] Shared interests can be discerned, allowing the niche to encompass and address the apprehensions of other stakeholders when formulating their approach to framing sustainability challenges (Smith & Raven, 2012, p. 1031). Freelance designers are ideal candidates to lead the developments of networks, taking knowledge and experience gained from working in one product category to the next client.

Fostering personal relationships with senior stakeholder representatives can prove particularly advantageous in fostering these networks. Quitllet emphasized the importance of his relationship with the owner of manufacturer Vondom in developing his designs during our interview:

> He [Pepe Jose Albania, the owner of Vondom] is going to put a lot of … him into the project and that's nice because he's not just a businessman who's going to say, 'Okay this can sell we do it, but it's my team who's going to work I'm just making the business'. He likes to be involved in the process (like Kartell also love to) and that says something. To create things that really respect the original thought, the original intensity from the beginning to the end. You need to have this kind of relation and you can't just send the drawing and say, 'Okay. I'll just put my name on it'.

These comments underscore the pivotal role played by the social dimension of social-technical transitions (Diaz et al., 2013, p. 62). Quitllet was not the only designer I interviewed to emphasize the significance of fostering close personal relationships with manufacturers. Starck enjoys the advantage of a long-established personal relationship with Claudio Luti, CEO of Kartell. This enduring association has played a pivotal role in Stack's successful collaboration with the company, resulting in many designs that have achieved outstanding commercial success.

The *Bell Chair* was the result of a two-year collaboration between Grcic, Magis and the engineering team at a third-party injection-moulding company. Magis (including owner and CEO, Alberto Perazza) had previously established and developed relationships with both the designer and engineering company over many years as part of a network of relationships with suppliers and mould makers across northern Italy. Grcic attributes the success of the *Bell Chair* project to a deep engagement with the engineering team, which finessed the development of a lightweight design using a new recycled material.

> Most chairs are developed in this kind of dialogue with engineering but never to that extent in my own experience … I was really excited about

this because I enjoyed the dialogue with this guy. I felt I was learning so much and … [working in] a very kind of natural, synchronized way … because you know all his rules or where he said, 'We can't do this'. We [disagreed] and then found a way forward rather than [making] a bad compromise.

While the importance and benefits of such a relationship appears obvious, they rarely occur. Grcic reported that he had never previously been so closely involved with the production of a product. As we have already seen, Grcic was also involved with the development of messaging and marketing collateral used to promote the *Bell Chair*. Although acting as a consultant rather than an employee, Grcic enjoyed the rare opportunity to become deeply involved with all aspects of design for this project.

Efficient and effective communications among all the professional disciplines engaged in a project are imperative to ensure timely production. However, maintaining effective communications is challenging, as different actors are focused on different priorities and exchange ideas using specific terminology. Design academic Tom Fisher explained the different priorities facing designers and engineers when work working on a project involving plastics, summarizing:

> While design engineers are concerned with physical, mechanical performance, product designers think of plastic in terms of consumers' engagement with objects through the tactility of their surfaces and the visual effects of their forms – aesthetic relationships with objects activated by culture.
>
> (T. Fisher, 2015, p. 123)

Fishers' comments further emphasize the benefit of the close working relationships that Grcic experienced. However, designers face specific communications challenges as they work with engineers and specialists to develop their creations for production, especially as actors are often in separate geographic locations, with language barriers adding further complexity.[9]

Research undertaken by Alaa Anssary examines the range of challenges to effective communication between designers and engineers including 'perceptual gaps, the use of different languages, and a lack of tools to describe the interplay between material attributes, shaping techniques and form aspects' (Anssary, 2006, p. x). Anssary found significant differences in the terminology used by designers and engineers. Designers preferred 'soft vocabulary based on their aesthetic experiences', while engineers used 'hard terminology based on their technical experiences' (Anssary, 2006, p. 139). The different vocabulary used to describe the same attributes can create confusion between designers and engineers when developing a project. Designers participating in this research went on to report they regarded engineers as patronizing, unimaginative and inflexible, further highlighting the difficulties in maintaining cordial relations between those actors (Anssary, 2006, p. 132).

Transition theories such as MLP are reliant on the process of experimentation for more sustainable solutions to emerge. Groups of actors work together to develop niche experiments which are evaluated by the market and other industry participants. Competitors benefit from this experience to guide their own projects, which might incorporate or adapt solutions from previous experiments or introduce a radically new approach. The practice of trial and error is accelerated by observing the success and failure of experimentation by other industry participants.

This process of experimentation can be observed by tracing developments in the plastic chair market. The first chairs to be designed using renewable plastic were produced by independent designer/makers with access to their own production facilities (Tom Price, Marcel Garcia, Dirk Vander Kooij and Joris Laarman) or from designers working with smaller manufacturers (A Lot of Brasil, Alki, DesignByThem, Hay, Label/Breed, Mobles114 and Moroso).[10] Niche innovations like these are crucial to initiate transition. As Smith observed, 'mainstream' actors only become interested in new technologies when niches have been proven 'in terms of scope for profitable application' (Smith, 2003, p. 1).

To significantly affect the overall market the support of larger manufacturers with established distribution networks is needed. But even for a designer with a track record of commercial success, finding manufacturers with a demonstrable commitment to sustainability goals remains challenging. Large incumbent manufacturers are often locked in by their investments, with established supplier networks (with their own lock-in issues) adding pressure to continue business as usual and defend the status quo. Sustainability is not a top priority for many manufacturers but the situation is changing fast.

Even when the support of a sympathetic major product manufacturer is secured, renewable carbon initiatives are susceptible to the aggressive defence tactics deployed by the petrochemical regime. These tactics include lobbying policy makers, influencing public opinion, and safeguarding advantageous technical standards as discussed in the previous chapter. Despite this formidable resistance, some designers and manufacturers have succeeded in forming a direct relationship with more progressive sectors of the petrochemical industry. In Brazil, retailer Tramontina worked closely with local petrochemical giant Braskem to develop two ranges of chairs. That partnership resulted in showcase projects for a new post-consumer recycled resin (including the *Sissi Chair* [Figure 4.9]) and a 100 per cent bio-based polymer made form ethane derived from sugar cane, *Jet* (Figure 8.10). Based in Brazil, Braskem is in a unique position having benefited from government supported development of the biofuel market, giving access to decades of research in processing biomass (De Oliveira & Coelho, 2017).

Petrochemical companies are coming under increasing pressure to improve their sustainability performance across all areas of activity. Maegerlein reported that in his work at BASF every client meeting now includes questions about sustainable plastics:[11] 'It started with the designers, absolutely. But [now] it's, I think every engineer, every company is looking desperately for solutions.' Emeco have enhanced their reputation for sustainable designs by working closely with BASF to develop new recycled materials. This programme of material innovation has helped attract leading designers such as Starck, Morrison and Barber Osgerby to work on Emeco projects. BASF also developed the water-based acrylic resin used by Aisslinger for the *Hemp Chair*, among the best-scoring designs included in my

Figure 8.10 Tramontina retail *Jet* manufactured using Braskem's 'I'm Green' bio-PE made from ethane derived from sugar cane. Courtesy Tramontina.

study. Designfabrik acts as a hybrid actor, bonded to the existing petrochemical regime but sympathetic to the sustainability challenges observed by other actors (Smith, 2007, p. 436). These direct interactions between niches and entrenched regimes are a key process in the take-off stage of transition (Smith, 2007, p. 427).

Material manufacturers can be active participants in sharing knowledge, transferring lessons from one project to the next. For instance, Shell was involved in most projects with solar photovoltaics in the Netherlands in the late 1990s, speeding up development of that industry (Schot & Geels, 2008, p. 544). BASF and Braskem appear to be playing similar roles in advancing renewable plastics. By helping to build social networks, material manufacturers can facilitate interactions between relevant stakeholders, and provide access to the necessary resources (people, expertise and finance) to accelerate transitions (Schot & Geels, 2008, p. 540).

e. Timing

The speed with which the transition to renewable plastics occurs depends largely on the participation of major manufacturers. Timing is everything: if they wait too long, they risk losing market share; if they go too soon, the risks are equally high. Philippe Starck told to me that he is working with Kartell to transform the company from its dependence on fossil plastics to a 'smart materials company'. He went on to explain that Kartell is working with bioplastics, responsible plywood and other materials made with 'human intelligence' in efforts to develop more sustainable products. Those statements represent a demonstration of the potential powerful agency held by organizations in guiding sustainable transitions when setting strategy. While designers are unlikely to hold much influence over the strategic policy settings of a large manufacturer, Starck's comments illustrate the important role that designers can play in guiding, developing and delivering sustainability commitments.

Starck's *AI Chair* (2019) represents Kartell's first attempt at incorporating recycled plastic into its repertoire, but the company has yet to include a bioplastic chair in its catalogue. Kartell did develop a prototype for a bioplastic chair in 2018 (*Bio Chair* by Antonio Citterio) using a polyhydroxyalkanoate (PHA), but this version of the design failed to make it into production. In 2021, the design was launched featuring recycled plastic (Figure 8.11). In 2019, a bioplastic version of its *Componibili* storage system was released, using the same PHA derived from 'non-GMO agricultural waste not intended for the food chain'.[12]

Figure 8.11 *RE* by Antonio Citterio for Kartell, 2021. Courtesy Kartell.

This pivot from fossil plastics was probably primarily motivated by falling sales, with company revenue estimated at €83 million in 2019 down from a high of €107 million in 2016, a 22 per cent decrease (Statista, 2022). At that time Kartell's product catalogue consisted almost entirely of fossil plastic products, exposing it to changing attitudes towards plastic. Certainly, the company's position has changed significantly from 2014 when, reflecting on fifteen years of success with its transparent PC range, all efforts toward dematerialization were abandoned. A press release announced that, following the introduction of *La Marie* (at 3.5 kg), 'minimalism was no longer necessary'. *Louis Ghost* weighed 4.8 kilograms and the company proudly boasted that even larger and heavier products using PC had been created, including the 18 kilograms *Ghost Buster* (2010) and culminating in the 30 kilograms *Uncle Jack* sofa in 2014 (Figure 8.12), all designed by Starck.

Fast forward three years and, in December 2017, Kartell, through parent company Felofin, backed a new strategy to embrace renewable plastics with a €10 million investment to acquire 2 per cent of an Italian bioplastic manufacturer (Bio-on, 2017). Bio-on, the first industrial-scale producer of PHAs was a supplier to Kartell and Unilever was awarded best bioplastic company of the year by an industry trade magazine in 2019 (Barrett, 2019b). The announcement of the deal between Kartell and Bio-on signalled a potentially productive alliance between an incumbent in the furniture industry and a new entrant to the bioplastics market,

Figure 8.12 *Uncle Jack* sofa, Philippe Starck for Kartell, 2014. Weighing almost 30 kg and 'the largest piece of transparent polycarbonate ever injected in a single mould'. Courtesy Kartell.

with the potential to accelerate Kartell's transition plans. However, at the end of that same year, Bio-on became involved in a financial scandal, following the publication of an unfavourable report from a financial analyst in Israel (who Bio-on accused of shorting the stock) (Reuters, 2019). The chairman/CEO of Bio-on was briefly arrested and the company declared bankrupt at the end of 2019 (Bioplastics magazine, 2020). Bio-on was eventually sold for a fraction of the value implied by Kartell's investment, making it unlikely the company will recover any funds.[13]

Kartell's experience to date illustrates the risks associated with experimenting with embryonic technologies. No guarantees can be given that such ambitious technical reorientations will succeed. Or, as Peen and Geels put it, 'the possibility of "betting on the wrong horse" makes green reorientation a risky process with long-term strategic ramifications' (Penna & Geels, 2015, p. 1029). Kartell's experience also illustrates that investing resources in technologies undermining powerful regimes can be risky.

Given the investment of time and capital required to adapt, it is more logical for large product manufacturers to reconfigure to adapt to the changing market (F. W. Geels & Schot, 2007). A refiguration strategy includes selecting and adopting symbiotic components from new technologies and making minor adjustments to established business practices to accommodate them. That approach maximizes the potential to defend market share, while minimizing the risks from working with relatively untested technologies (F. W. Geels & Schot, 2007, p. 410; Smink et al., 2015, p. 88; van Mossel et al., 2018, p. 57). Kartell's experiment with recycled plastics (the *AI Chair*) is more aligned to this shadow-track strategy. Both Vitra and IKEA have also introduced chairs made from recycled plastic, but neither has yet produced a bioplastic design. Whether these innovations represent the beginnings of the transition to renewable carbon materials by these major manufacturers or are merely designed to mitigate tension among critics and pre-empt the imposition of additional environmental regulations remains to be seen (Diaz et al., 2013, p. 67). Certainly, Kartell has since re-released several of its most popular designs using recycled plastics and Vitra have announced plans to follow suit.

It is probable that other major manufacturers will explore similar cost-effective and less precarious opportunities compatible with existing investments when diversifying into a new niche. This process of adaptive diversification is a less disruptive hedging strategy, allowing current business activities to continue as usual while experimenting with promising innovations (van Mossel et al., 2018, p. 57).

Among the major manufacturers IKEA has also made a commitment to transition away from fossil plastic. IKEA aim to 'begin to phase out virgin plastic from the IKEA product range, a focus that will continue towards 2030' (IKEA, 2020, p. 15). Although commendable, it should be noted that this statement makes no firm commitment to reducing the use of virgin polymers. It is a disappointing departure from IKEA's 2020 sustainability goals (2012, revised 2014) that committed to 'all plastic material used in our home furnishing products will be 100% renewable and/or recycled' by August 2020 (IKEA, 2020, p. 14). At the start of 2024 the IKEA catalogue featured only one chair made from renewable plastics. IKEA, together with all the major established manufacturers, risk losing market share if consumer sentiment in favour of renewable plastics accelerates faster than their transition plans.

2. Summary

Hybridization offers a proven pathway for designers and manufacturers to minimize the costs of transition, as established manufacturing processes can remain unchanged. While purists may question the validity of hybridization as a solution for plastics from an environmental perspective, the reality is that most recycled plastics available are already hybridized. Virgin plastics are commonly combined with recycled plastics to help maintain structural integrity. However, the petrochemical regime's emerging preference for the mass balance approach to recycling (where minimal amounts of recyclates are included in materials certified as being recycled) will likely continue to be a topic of debate. Hybridization is also a popular strategy to imbue bio-based plastics with the strength often needed for consumer goods. The incorporation of bio-based plastic into a product can negatively impact its end-of-life prospects. However, the inclusion of any proportion of biomass in the manufacturing process diminishes demand for fossil fuels, particularly pertinent for products intended for prolonged usage, where end-of-life impacts are delayed.

When investigating renewable carbon materials designers can familiarize themselves with the story of their creation. Recycled plastics carry the history of their previous lives, while bioplastics are derived from an increasing variety of feedstocks, each with their own unique provenance, allowing engaging marketing narratives to be constructed. These factors differentiate renewable plastics from fossil plastics, created through extractivism and exploitation. By educating marketing teams on the provenance of materials, designers can introduce a unique and compelling differentiator in favour of renewable plastics in sales and marketing collateral.

Designers can play an important role in tailoring their products for the specific needs of market segments most likely to be willing to pay a premium for environmentally friendly products. The complexities of identifying and targeting environmentally conscious consumers as a discreet market are acknowledged. Targeting niche markets with potential for significant growth is a proven strategy to accelerate transitions. In coming years, governments (at all levels) can be expected to update and revise purchasing policies, with greater emphasis given to the importance of sustainability considerations, encouraging or even mandating the purchase of more sustainable products. When such changes occur, renewable carbon-based products, including chairs, will benefit from significantly increased demand, rewarding innovative manufacturers with compliant products already available.

Combining multi-technological solutions is a proven strategy to accelerate transitions generally. Many renewable plastics are suitable for experimentation with new digital manufacturing technologies or processing with technologies borrowed from other industries. The success of the innovative designs discussed in this chapter demonstrate that this approach can result in the development of novel design solutions, while promoting the update of renewable plastics.

While the focus of this chapter is on the role of the designer, an exploration of their agency within the broader network of actors engaged with all aspects of production is necessary to understand how regime change can occur. Establishing networks of like-minded actors to explore and share experiences from individual niche experiments is essential. Several of the successful designs included in this study were created by designers with well-established relationships with senior-level personnel representing product manufacturers. Support from material manufacturers with vested interests in fossil fuel extractivist activities is more challenging to secure. Large incumbent petrochemical organizations, with access to their own fossil feedstocks, are unlikely to voluntarily switch to renewable organic raw materials supplied by third parties, thereby diminishing control of supply chains (Bennett, 2012b, p. 106). Despite those barriers exceptions can be found, with certain material manufacturers embracing (or at least experimenting with) transitional technologies, as exemplified by Braskem. Furthermore, these are joined by major petrochemical organizations which are not directly owned by those involved with fossil fuel extraction, such as BASF, which has also pursued a dualistic approach.

Cultivating relationships with key material manufacturers such as these, who are prepared to reconfigure their operations to accommodate new feedstocks, will provide access to expertise developed from the real-life experiences of other pioneering clients. Indeed, fostering close relationships between entities operating on all three levels of the MLP model is the most productive way to accelerate change, as it nurtures a process of co-evolution and mutual adaptation (Schot & Geels, 2008, p. 547).

Finally, Kartell's failed investment in a bioplastic manufacturer illustrates the financial risks in backing new, commercially unproven technologies. In contrast, competing large-scale manufacturers have been more cautious in their approach, launching a limited range of designs made using recycled plastics and adopting a 'wait and see' approach, particularly in the case of bioplastics. Many smaller manufacturers are introducing novel chair designs that incorporate either recycled plastics or bioplastics, and this trend is expected to continue for the foreseeable future across many categories of consumer goods. With major manufacturers unlikely to surrender market share, their transition to renewable plastics can be expected to accelerate in the coming years. Nonetheless, this transition is not guaranteed, as it contingent on a multitude of variables including mainstream consumer acceptance of renewable plastics and the co-evolution of the petrochemical regime to accommodate renewable feedstocks. Both designers and manufacturers wield considerable influence in shaping the search for another way 'to be', by actively advocating for the adoption of renewable plastics.

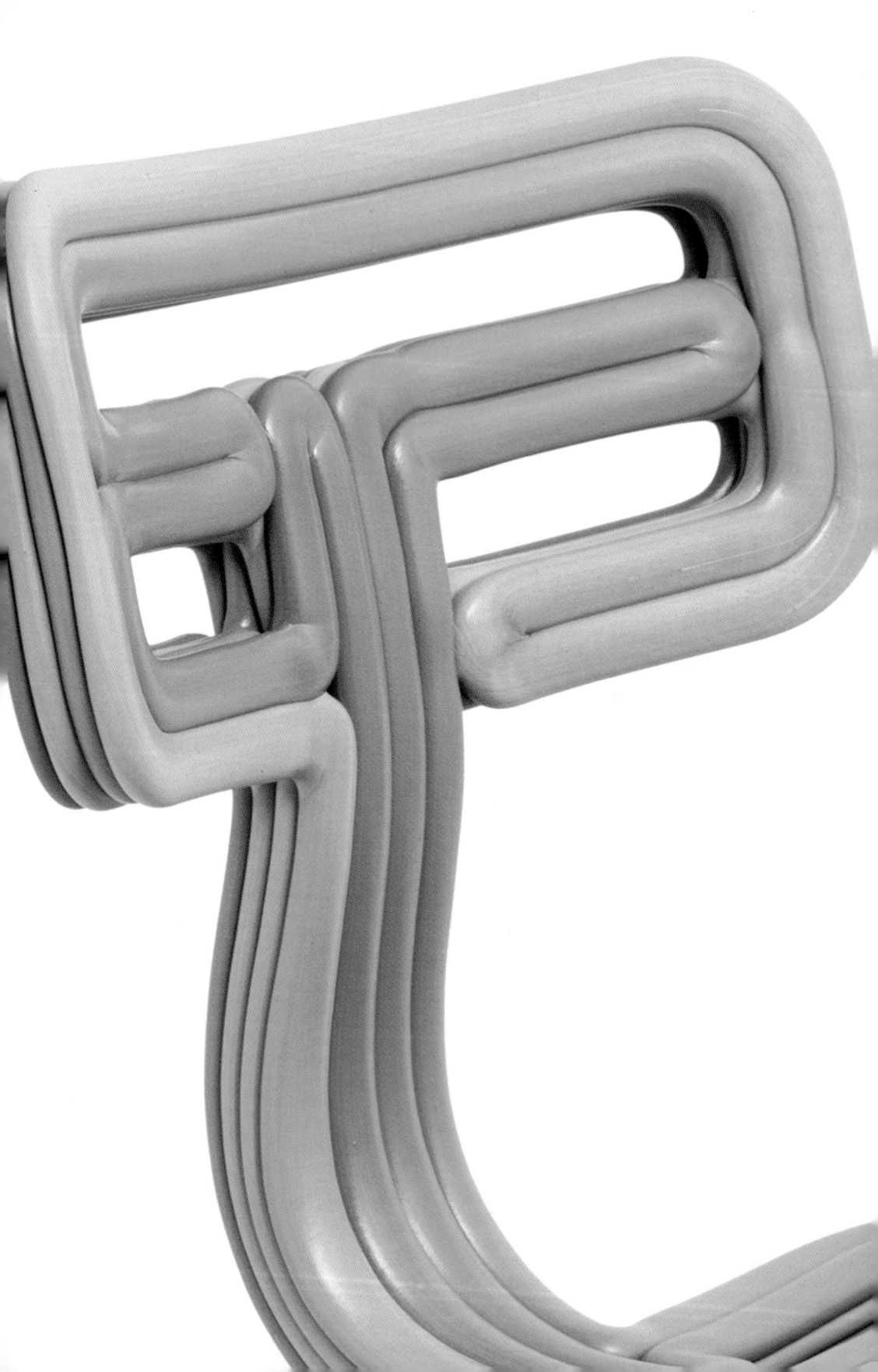

Conclusion

Are you really sure that this world needs another plastic chair?

During my interview with BASF's Andreas Maegerlein, I was surprised to hear him suggest that this question should be asked of every designer proposing another plastic chair. Maegerlein shared that his company is coming under increasing pressure from clients and governments alike to provide solutions to the plastics issue, predicting:

> There is a big change coming because, people will not accept the materials [plastics] as they were anymore … companies are now seeing that they have to act before their products will be … taken from the market due to restrictions coming from politics. They have to find solutions … It's a stupid idea to tell people to use less of our products but actually, that is the way you need to think.

While I found Maegerlein's comments surprising, they are also encouraging, with a representative from a prominent member of the entrenched petrochemical regime acknowledging that the operating environment is being disrupted and they are endeavouring to adapt their business in response. This approach contrasts with the disappointing level of commitment to sustainability initiatives among many of the product manufacturers selected for the case studies in this book.

The operating environment for the petrochemical regime continues to be impacted by significant disruptions detailed in the Chapter 6, such as the war in Europe and China Sword. The issue of ocean plastics has remained in the public discourse, exacerbated by the realization that an estimated 6 billion face masks were being dumped in our oceans every month during the height of the Covid-19 pandemic.[1]

The landscape for the entire fossil fuels regime is also facing disruption. As at the start of 2024, 2,355 jurisdictions across forty countries – representing over 1 billion citizens – had declared a climate emergency.[2] Chaos reigns in energy markets, reflecting turmoil in the fossil fuel markets caused by Covid-19 and the ongoing war in Europe. As more people accept the science of climate change, it is inevitable that all extractivist activities will be held increasingly accountable for the damage they are inflicting on our environment. Pressure for change aimed at both the petrochemical and fossil fuel regimes will intensify. Multi-level perspective (MLP) theory predicts that significant disruptions such as these can create an opportunity for new technologies to challenge the entrenched regime.

Despite favourable conditions, sustainable transitions take decades (Elzen et al., 2004, p. 108; F. W. Geels, 2010, p. 495; Markard et al., 2012, p. 956). Transitions are 'enacted by interpretive actors that fight, negotiate, search, learn, and build coalitions' (F. W. Geels, 2010, p. 495). A societal transition requires far-reaching changes across multiple dimensions: technological, material, organizational, institutional, political, economic and socio-cultural (Markard et al., 2012, p. 956). A single event or person cannot cause transition – it is a process requiring government intervention, shifts in consumer attitudes and large-scale investment in new technologies (Elzen, 2004, p. 108). The shift to renewable plastics will require enormous investment in mechanical, chemical/advanced recycling infrastructure, and the research and development needed to produce bioplastics at industrial scale.[3] However, supply (and the transition investment required) will not increase unless matched or exceeded by demand. The purpose of this book has been to identify practical actor-specific tasks that supportive designers and manufacturers can undertake to accelerate the transition to renewable plastics.

In the twenty-first century, renewable plastics potentially enable us to continue to enjoy the many benefits offered by plastics, while reducing the environmental devastation caused by their creation and disposal. Plastics are often combined with other materials when manufacturing products, making it challenging to accurately calculate their environmental impacts. In response, I developed a simplified eco-audit tool (Environmentally Responsible Product Rating or ERPR) to facilitate a comparison between designs using only the very limited information commonly available from chair manufacturers. One of the main innovations of the tool is an attempt to quantify the appeal of a product while evaluating its environmental impact. A 'sustainable' product only delivers an environmental benefit if it is purchased in preference to a less-sustainable alternative, a fact which is often overlooked by methodologies focused on measuring environmental impacts. For a product to sell it must appeal to end-users, the role of design is crucial here. While the EPRP tool was useful to identify successful product level initiatives, my main objective is to examine how the knowledge gained from experiments with renewable plastics can be scaled-out and scaled-up to drive change across consumer goods at the socio-technical level.

This study found that specifying renewable plastics alone is not sufficient to optimize a design for sustainability. The most successful designs identified by the ERPR tool, embrace the advances in dematerialization achieved during the first seventy years of plastic chair development (Chapters 1 and 2). Designers were often prepared to consider manufacturing technologies both new and old. Some of the chairs with the least environmental impacts were developed using low-pressure manufacturing technologies. Combining multiple technologies is a proven strategy to accelerate transitions as shown by MLP studies and discussed in Chapter 8. Thermoforming and compression moulding are more widely accessible, due to the lower upfront investment needed, making smaller production runs economically feasible. These processes operate at lower temperatures and require lower operating pressures, thereby avoiding the cost of expensive steel moulds while consuming less energy than injection moulding. Additive manufacturing technologies are rapidly maturing and will be of increasing interest to designers and manufacturers working with renewable plastics. Even small organizations, operating in tiny or remote markets, can experiment with renewable plastics using these accessible manufacturing technologies.

Tailoring products to meet the needs of a well-defined target audience is a proven successful marketing strategy. Further, MLP posits identification of potential high-growth niche markets as a strategy to accelerate transitions (Chapter 8). More rigorous attention to researching and satisfying the requirements of environmentally conscious niche markets could lower the financial risks of developing new designs, while accelerating the transition toward renewable plastics.

My comparison of the environmental impact of designs also found that improving circularity, by using recycled plastics, is currently more likely to deliver chairs with superior environmental performance. Many of the bioplastic chairs included in this study failed to achieve above-average ERPR scores (Chapter 4). However, recycling alone cannot provide the solution to the plastics crisis, with bioplastics offering the only permanent alternative to our dependency on fossil fuels. Future experimentation should not be deterred by the disappointing scores awarded to the bioplastic chairs included in this study, as the pursuit of perfection necessitates ongoing applied research. Early attempts at incorporating bioplastics into consumer goods are useful to alert material manufacturers to the challenges that need to be addressed to meet the expectations of designers and product manufacturers.

Experimentation and learning by doing is an essential part of transition. Nora Fehlbaum, CEO of Vitra explains, 'things need to be tested, mistakes must be made, and in the process the company might sometimes overlook an important aspect or underestimate the impact of an activity' (Bickersteth, 2024). Niche accumulation is the very process of networking and learning from the experiments undertaken by innovators in the industry over the past fifteen years. Only by examining them can the most 'promising next steps' be identified (Elzen, 2004, p. 290). For example, at the 2023 *Salone del Mobile*, Prowl Studio showcased a prototype of a flat-packed *PEEL* made from PLA reinforced with hemp (Figure 9.1). Chairs need to endure significant mechanical stress, and this design demonstrated that the material could meet these demands, regardless of its biodegradable credentials. Transitioning to a more sustainable long-term solution requires designers and manufacturers to continue experimenting with bioplastics to reduce our dependence on fossil fuels while retaining the unique advantages offered by plastics.

Promoting the use of renewable plastics is important, as it is the only short-term activity that can reduce the GHGs emitted during their production, as shown in Chapter 3. Stimulating demand from the consumer goods sector extends the use of renewable plastics from the packaging industry, where they are already established. Developing new market segments for these materials creates additional demand, incentivizing material manufacturers to invest and increase supply. In turn, this maximizes returns, as new economies of scale are achieved from

Figure 9.1 *PEEL*, Prowl Studio, 2023. Stacking chair made from hemp (including hemp foam and hemp bio leather) that can be industrially composted. Courtesy Noah Webb for Prowl Studio.

servicing a larger market. Consumer goods represent an especially appealing niche for material manufacturers, as profit margins are potentially considerably higher when providing the superior quality, durable materials aligned with the standards sought by this sector.

A modest drop in demand for virgin plastics has the potential to cause significant disruption to the fossil regime as discussed in Chapter 3. The significance of a relatively small change in demand for fossil plastics highlights the importance of hybridization as a strategy to advance the uptake of renewable plastics. While not a perfect solution, hybridization allows current work practices to continue and reduces demand for fossil fuels. If a quarter of the virgin plastic used to make consumer goods is replaced with renewable plastics this would be sufficient to achieve the destabilizing reduction in demand identified by Carbon Tracker. Even if this target proves too ambitious, the study by Zheng and Suh (also discussed in Chapter 3) suggests that reducing the CAGR of virgin plastics to 2 per cent could be sufficient to achieve emission targets if demand management efforts are combined with the other strategies they outline. Action by those involved with the production of consumer goods is crucial, as bans on SUPs by the G20 countries, together with the introduction of EPR schemes, and a virgin resin tax will not be sufficient to curb growth in demand for virgin plastic in those jurisdictions (Back to Blue, 2023, p. 4). Overall demand will continue to increase but at a lower rate if only these three measures are implemented. Action by designers and manufacturers can add to this downward pressure on demand. This study demonstrates that designers and manufacturers of consumer goods can play a crucial role in accelerating the transition away from our finite fossil resources.

1. Checklist for working with plastics

The responsibility of the designer to consider sustainability has garnered considerable attention from both scholars and practitioners ever since Victor Papanek advocated for the profession to be held responsible for its environmental and societal impacts. More recently, designer, and academic Rachel Egenhoefer bluntly stated:

> If product design lays the basis for the formation of materials, objects, services and systems, then the product designer's influence over the sustainability of production and consumption is nothing short of pivotal.
>
> (Egenhoefer, 2019, p. 99)

Design academic Vicky Lofthouse is among an increasing number of voices going further:

> Designers need to take responsibility for what they agree to design and say 'no' to designing nonsensical products that society really does not need.
>
> (Egenhoefer, 2019, p. 113)

Tony Fry argues that once the defuturing caused by the existing regime is exposed, failure to act 'is to commit oneself to remaining unknowingly within the anthropocentric and productivist structures of unsustainability' (Fry, 2020, p. 126). Such demands are easy to make by those with secure incomes but pose a greater challenge to those managing studios, who have responsibilities to employees and their families with livelihoods dependent on continued commercial success.

Adopting a more practical approach, Stefan Diez updated a checklist (Table 9.1) originally developed by Dieter Rams in 1976 as a guide to 'good' design, with a

Table 9.1 Stefan Diez's update of Deiter Rams guidelines for circular design

Circular Design Guidelines by Stefan Diez An update on Dieter Rams ten guidelines for good design developed in 1976
1. A good product remains useful for a long time.
Design it in such a way that it can adapt to changing requirements and thus remain relevant for longer.
2. A good product is repairable.
Use the kind of materials where signs of wear do not diminish the product's value. Construct it in such a way that components with a shorter life cycle can be replaced by the customers themselves.
3. Can the product be designed as a system?
Then, system modules or assemblies can be continually advanced and optimized by the manufacturer in line with technical progress. A good product can be updated and remains on the market for a long time.
4. Use materials that originate from a material cycle or that are renewable.
The materials used should not evaporate or rub off during use and should in general not be toxic. The materials used should be self-explanatory and easily separable from one another. A recycling point should be easily accessible to the last user.
5. As little energy as possible should be used in the manufacture, use and recycling of the product.
Consider the consumption of energy and resources over the entire life cycle of the product. In the case of products for everyday use, a high level of effort in production can be more than offset by daily savings. Keep the energy input for recycling in focus.

6. Design the product so that it can be transported in a space-saving way.
It can be packed in a compact fashion during production, for transport to the customer, for moving, for repair, and for recycling. The packaging provides reliable protection for the product against damage. Transport should generally be kept as short as possible.
7. A good product is innovative and fascinating.
It may be complex, but never complicated. And it rewards its user with a real benefit. Products should be coherent and honest throughout, speak for themselves, and enable a resonant relationship with their users.
8. A good product is used by many.
It can potentially be rented, shared and returned. Could the product, or a substantial part of it, remain the property of the manufacturer, who sells only its use? Maintenance and repair would then be part of the manufacturer's promise.
9. The production, maintenance & recycling processes employ people in a fulfilling and demanding way.
Good products are made in countries that ensure equal treatment for minorities and guarantee freedom of speech. The health of workers is protected. Workers are employed according to their skills and receive fair pay.
10. A good product is as little product as possible.
It consists of as little material as necessary or has even been replaced by a (digital) service.

Adapted from the Diez Office website www.diezoffice.com/circular-design-guidlines (accessed 5 March 2024).

focus on sustainability, which serves as a concise summary of considerations to be reviewed by those creating another chair, or any consumer good incorporating plastics. Numbered lists risk being interpreted as suggesting a hierarchy, with topics most worthy of attention appearing at the top. It is not clear if this is the intent, certainly all the items on the list are essential concerns. However, when designing using plastics a sub-set of items on this list, categorized under four headings below, require below particular attention. The last item on the list, 'a good product is as little product as possible', is the most important consideration for most consumer goods and particularly important for products primarily made from plastic.

a. Dematerialization

When either plastic or renewable plastic is used to make consumer goods that do not consume resources during their use the energy consumed in creating the material will cause the biggest environmental impact (Chapter 4). Any reduction in material usage leads to a decrease in GHG emissions across the entire life cycle, encompassing manufacture, transportation and end-of-life stages, addressing item 5 on Diez's list. The environmental impacts generated by the distribution network can be further minimized by optimizing products to reduce spatial demands during shipping and storage (item 6).

During the second half of the twentieth century, the interests of environmentally conscious designers and industry were often aligned. By reducing the amount of energy and/or materials needed to make products, including chairs, both the environmental impacts and the cost of production of individual units were reduced. As detailed in Chapters 1 and 2, those dematerialization efforts were initially rewarding; however, continued growth in demand quickly outweighed diminishing returns. Importantly, dematerialization lessons from this period are still valid and should be applied to lower the environmental impacts of consumer goods at the product level.

While dematerialization efforts traditionally focus on reducing energy consumed during production this should be broadened to include the minimization of impacts across the energy supply chain. Switching to renewable energy could significantly reduce the GHG emissions created during the production of fossil plastic by up to 62 per cent in 2050, which is a far larger reduction than could be achieved by switching feedstock to renewable carbon sources (Zheng & Suh, 2019, p. 375). Details of energy sources used to manufacture plastics are not often disclosed. However, just by raising queries about energy sources designers and manufacturers are signalling the existence of a potentially lucrative market for plastic made with renewable energy, while adding pressure to material manufacturers to consider this aspect of their GHG impacts. Driving this change is particularly important as the switch to renewable energy is a long-term project likely to encounter significant resistance from the entrenched petrochemical regime. Change is beginning to happen, with some material suppliers already using some renewable power.[4] This transition can be accelerated by supporting the material manufacturers that already use renewable energy.[5]

Furniture manufacturers and their supply chain partners should also be encouraged to use renewable energy. The final decision on energy supply rests with the entity consuming and paying for the resource. Designers can attempt to influence those decisions by actively championing the potential marketing advantage stemming from the transition to renewable power, by promoting the reduced environmental footprints of products. Among the furniture manufacturers included in this study S.CAB are noteworthy here, as they already use certified recycled polymers produced using renewable energy for their GoGreen series of chairs. Also already recognizing this potential, Basque manufacturer Alki are building a 'zero energy' production facility to be powered exclusively by renewable energy.

All these efforts can be considered an extension of dematerialization, reducing the resources consumed to manufacture a product. Designers are uniquely placed to represent the shifting priorities of the market, advocating for environmental improvements across all the inputs needed to manufacture their creations. Switching, at least partially, to some renewable energy is a relatively simple and cost-effective strategy available to many manufacturers and their suppliers.

b. Think virgin last: Avoid virgin plastics

After exhausting attempts to reduce the quantity of plastic required designers should focus their efforts on avoiding, or at least minimizing, the use of virgin plastics (item 4 on Diez's list). With a continually expanding and diverse range of recycled plastics and bioplastics available, including hybridized bio-based materials, the likelihood of finding a suitable option for the specific task at hand continually increases.

Product designers, while not environmental scientists or material engineers, have the expertise and experience to play a crucial role in the material selection process by advocating for renewable plastics. Designers can help identify appropriate sustainable materials that fulfil the requisite physical properties while effectively communicating the environmental advantages to their clients. Keeping informed of new materials is challenging. Designers interviewed for this study reported difficulty in accessing timely, accurate and impartial information on renewable carbon materials and their availability. No widely recognized sources were cited, with participants frequently resorting to ad hoc methods to find relevant information on renewable carbon materials and supplier details. Maintaining robust industry contacts is crucial to stay up to date on developments and supply issues.[6]

Effectively discouraging or delegitimizing the use of virgin plastics can only be achieved when viable alternative materials are readily accessible. Nominating a choice of renewable plastic will be partly determined by their availability, which will vary by region reflecting local conditions. Bioplastics are currently often only available from small-scale, geographically dispersed producers, each with their own product offerings reflecting local conditions. Recycled plastics are more likely to be available where suitable infrastructure is concentrated. Bioplastics are more likely to be available where suitable corps are abundant, or in regions that have access to the resources needed to invest in developing advanced generation feedstocks, such as algae or CO_2. Recognition of this regional variation in feedstock availability is reflected in the United Nations Environment Programme, which calls for technology to be adopted when it is compatible with the environment and society in which it is used (Holtz et al., 2008). This recognizes that nations or regions must find their own way of combining technological change within the confines of local conditions. While this approach appears logical, it is challenging to implement in a globalized world addicted to the economies of scale delivered by centralized homogeneous mass production.

Many renewable plastics lack an established track record to demonstrate their durability, strength or other attributes required for a proposed task. Designers need to work with both material manufacturers and product manufacturers and often even rely on their own investigations to satisfy themselves that a material is suitable. More technical research is required to provide independently verifiable data on the physical properties of new renewable plastics to support designers in assessing their relevance for a specific task. More investment in developing bio-based alternatives to the additives and colouring agents is needed. Universities can play a pivotal role in addressing and filling these information gaps, by establishing communities of practice to facilitate and encourage the dissemination of relevant information between designers and manufacturers.

A singular solution to the question of what do use to replace virgin plastic has not yet been conclusively determined, and it remains challenging to nominate the most sustainable alternative(s). MLP studies demonstrate this is not unusual for an emerging technology, there are multiple possible transition pathways and, as the notion of sustainability remains contested, different actors will disagree about the most desirable innovations and transition pathways (Köhler et al., 2019, p. 3). Finding solutions will require the exploration of many blind alleys and it can be expected that solutions will be distributed rather than hierarchical. Critics of transition theories argue that stakeholders' visions of the future are inevitably shaped by the systems and social environments they inhabit (Shove & Walker, 2007, p. 766). Certainly, there is a growing body of evidence to demonstrate how the petrochemical industry is attempting to protect their interest in this transition (Chapter 7).

While large and mid-tier manufacturers remain predominantly locked-in by previous investments and pre-existing supply arrangements, designers wanting to work with innovative materials are more likely to find support from smaller manufacturers with a strong commitment to sustainability, combined with the ambition, networks and financial resources needed to support innovation. While that checklist may appear formidable, suitably qualified organizations do exist. In fact, many of the designs featured in this study have been developed by entities that match these criteria.

c. Design for a long and circular life

Designing products to withstand years or decades of use and abuse, postpones any end-of-life impacts and alleviates the need to buy replacement products (item 1 on Diez's list). If wear and tear or product upgrades are likely to require the replacement of specific components, products should be designed to facilitate ease of access for repair (item 2). Particular consideration should be given to how different materials or attached or bonded. For example, Luke Pearson and Tom Lloyd of Pearson Lloyd design studio observed that the *Egg* and *Swan Chairs* designed by Jacobsen and Saarinen's *Womb Chair* feature fabrics glued to concave padded shells, typical of mid-century design (Frearson, 2023). This solution makes the products easier to manufacture but renders recycling unviable. They argue that this upholstery solution is unacceptable today and suggest designers should give preference to linear or convex shapes, where textiles can be held in place with drawstrings rather than glue. In all cases, products should be designed for disassembly to facilitate repairs and, at the end of their useful life, promote circularity with plastic components permanently labelled with RICs to encourage recycling. As EPR initiatives become more common designers and manufactures will be forced to give more consideration to what happens to their products at the end of their useful life.

d. Promote the provenance of renewable plastics

Designers now need to take the lead in championing renewable plastics as the most sustainable and logical choice – an essential component of our response to the environmental emergency in the age of the Anthropocene. Design, in its broadest sense, has the responsibility and opportunity to guide cultural preferences towards renewable plastics, by highlighting their merits and advantages. Indeed, designers have previously influenced the cultural interpretation of the material. Plastics were originally admired for their formability and low cost, they represented the ideal expression of an abundant machine age, the ultimate signifier of modernity (Chapter 1). The material was revered as a scientific miracle, the enabling driving force fuelling consumerism while perpetuating the illusion of endless material progress. For product designers plastics enabled previously unrealized dramatic, organic, curvilinear and otherworldly shapes. Plastics were promoted by the creative community with promises to eliminate scarcity and create a brighter, shinier, easy to clean and more hygienic world.

By the 1970s, plastics were both venerated and derided for their disposability, malleability and homogeneity, having simultaneously created a utopian profusion of affordable consumer goods and an abundance of ephemeral waste (Chapter 2). Plastics are often considered as superior to natural materials by designers and manufacturers, primarily due to their reliable homogeneity. Nonetheless, poorly crafted plastic products frequently garnered disapproval, particularly when the material was inappropriately deployed as a low-cost substitute for natural materials. In the early 1990s, Manzini recognized the main challenge facing designers working with plastics was to shift the focus of consumer attention from quantity to quality (Manzini, 1992, p. 5). Plastics were increasingly recognized as shameless imitators, fake or phony, failing to deliver the modernist dream. Industry continued to focus on quantity, leading to plastics being increasingly recognized for their role in the destruction of our environment and threat to our health.

Ideological principles and community attitudes toward plastics have been moulded, shaped, influenced and shifted by the creative industries. Our cultural interpretation of plastic has been guided by product designers, interior designers, set designers, lighting designers, graphic designers, UX designers, architects, fashion and jewellery designers. Articulating positive cultural visions helps to legitimize innovations, attracting further support. Plastics are responsive to cultural construction. Contrast *Barbarella* (1968), starring Jane Fonda in a futuristic eroticized, plasticized world, with *Blade Runner* (1982), where the dystopian future featured a protagonist inhabiting an apartment furnished with wooden Mackintosh chairs and black leather sofas, notably devoid of the curvaceous, glossy, shiny utopia of the, by then, rejected Space Age. In 2023, a film based on a plastic doll became the biggest box office success of the year. These cinematic depictions have helped shape public perceptions of plastic and influenced how the material is perceived and used in society.

Renewable carbon materials offer no obvious consumer benefits compared with fossil plastics. In fact, perceived disadvantages are more common, especially the aesthetic limitations that afflict many renewable plastics. Revealing the detrimental origins of fossil plastics is a strategy that has the potential to stimulate demand for renewable plastics and increase their use in consumer goods. Designers should adjust their focus from 'truth to materials' to declaring the truth about materials. End-users must be alerted to the destructive defuturing consequences of their purchase decisions.

Designing products using virgin plastics perpetuates the incumbent regime and legitimizes their activities. As I write, multiple new fossil plastic manufacturing facilities are being constructed, often without consent from marginalized and vulnerable local communities, ignoring their rights and cultural traditions. Vast areas of land are being cleared, removing any sign of the natural environment. Once levelled, this land is used to build infrastructure dedicated to devouring enormous quantities of precious fossil fuels while spewing methane and other climate-destroying GHGs into our environment and threatening the health of the local community for decades to come. Fossil plastics are born out of violence and environmental destruction, traits inherited by the material and culminating in yet more devastation for our landscape and its inhabitants when discarded.

Highlighting the negative social and environmental impacts of virgin plastics should be approached with caution. Demonizing the material could be counterproductive, encouraging end-users to rid themselves of perfectly serviceable products only to replace them with similar products made from alternative more sustainable materials. The defuturing impact of past purchase decisions can only be made worse by accelerating the purchase of replacement products, regardless of how impressive their environmental credentials (Fry, 1999, p. 2).

Alternatively, the rich provenance of renewable plastics offers significant potential for those involved with the promotion of products (discussed in Chapter 8 and addressing item 7 on Diez's list). Initiating and developing a closer relationship with marketing to develop narratives is made particularly challenging when freelance designers are engaged on a project basis. There is a danger that marketing departments will likely remain oblivious to the significance of a material's provenance unless it is highlighted.

Promoting the provenance of renewable plastics can shift our cultural interpretation of the material. Highlighting the previous life and origins of these materials allows the development of interesting and engaging narratives to differentiate products made from renewable plastics. Of course, these claims must be backed by resilient products designed to withstand the vagaries of fashion in addition to daily use, and abuse.

The *Bell Chair* project sets the gold standard here, as detailed in Chapter 8. The consultant designer (Grcic) worked closely with the marketing department at Magis to construct a compelling narrative to explain where the material had come from, its history, and the environmental benefit of using it. This contrasts with the *Odger*, developed for IKEA, where information about the material or the environmental benefits of buying the chair are unavailable on the IKEA website. Without this background information a potential purchaser might struggle to rationalize the price premium while being asked to compromise on aesthetic choices (the *Odger* is only available in four colours and two variants).

Different strategies have already been developed by designers working with renewable plastics to overcome any perceived aesthetic or sensory deficiencies as discussed in Chapter 7. Colour choices are usually restricted and often muted and clichéd, especially so for post-consumer recyclates. Importantly, the heterogeneous nature of renewable plastics enables visually unique products to be created using mass production techniques. In addition to being visually distinct from virgin plastics renewable plastics often feel different to the touch, creating exciting possibilities for new seductive tactile interactions. Evangelizing the unique appeal of these new sensory experience can help differentiate products and drive sales, ensuring the environmental benefits promised by a product are delivered.

In the final analysis, the success and speed of any transition is determined at the micro level by the individual attitudes, motivations and individual purchase decisions. Designers can work with manufacturers to nudge these decisions through a tighter focus on developing the marketing narrative. Providing information to purchasers to help them to identify sustainable choices makes them active and vital participants in the transition. While it can be challenging to deliver complex messaging, I was surprised to find the industry is failing to take full advantage of recent innovations in digital technology which facilitate interaction with prospective purchasers. For example, QR (quick response) codes could be embossed on the underside of a product, added to labels or point-of-sale material as a simple, low-cost tool to directly link prospective purchasers to online content. This would allow for a detailed explanation of how products are designed and manufactured to improve environmental credentials using engaging, interactive content. Manufacturers could also benefit from savings with this strategy, relying less on training point-of-sale personnel to disseminate information and reducing waste and misinformation created by obsolete hardcopy collateral.

2. Duty of design

A constantly growing, increasingly urbanized and affluent global population will demand more manufacturing to support ever higher levels of consumption. This might continue to deliver sustained growth for the linear capitalist economy, but at the continued expense of our environment. Meeting emission reduction targets requires the petrochemical industry to end its dependence on virgin fossil fuels.

Transitions are shaped by the actions of multiple actors including regulators, interest groups, designers, manufacturers, academic researchers, retailers and consumers. Design is uniquely placed to help accelerate this particular transition. The main priorities for designers are to minimize their use of plastic (or any other material for that matter). Where plastics are still necessary efforts should be made to avoid virgin fossil plastics where possible by using renewable plastics. Designers also have a responsibility develop aesthetically appealing, long-lasting, repairable products and to encourage the use of renewable energy by manufacturers and their supply chains. Designers need to take responsibility for explaining the provenance of the materials they are using and evangelize the benefits of renewable plastics.

With consumption of plastic rapidly approaching half a billion tonnes per year the transition to more sustainable materials can only realistically be achieved by reconfiguring the existing petrochemical regime. As Geels and Turnheim observe, 'reconfiguring sociotechnical systems offers greater analytical traction and stronger socio-political appeal than the overhaul of capitalism or calls for frugality and sufficiency' (F. W. Geels & Turnheim, 2022, p. 13). Adapting material manufacturing processes at this scale will be challenging, but the industry has done it before. The North American shale oil boom only commenced in the mid-2000s and, in less than two decades of significant investment, already produces 2.7 million barrels of ethane every day. With the right policies, and the support of the multi-actor network discussed in Chapter 6, the industry can reconfigure equally as quickly to vastly reduce its dependence on virgin fossil fuels and replace them with recyclates. A review of transitions occurring in the electricity, transport and heating regimes in the UK found reconfiguration enacted by incumbent firms as central to the success of a transition, emphasizing the importance of this strategy in transitioning the petrochemical regime (F. W. Geels & Turnheim, 2022).

My analysis identifies a practical course of action for designers to embrace a more sustainable approach to the design of plastic products while operating in the current neoliberal world. The MLP model has been criticized for being representative of the neoliberal system 'in which scarcity is (mis) assumed to be observed in the market and the environment is seen as an endless resource and waste sink open for exploitation by our socio-technical system' (Gaziulusoy, 2010, p. 110). Tony Fry argues that recycling can be labelled an example of 'environmental technologies' that 'totally buys into technology's inscribed discourse of a solutions-based model of progress' (Fry, 2020, p. 212). Replacing fossil plastic with renewable plastic can be criticized as reformism or solutionism, assuming every social problem has a technological fix (Morozov, 2013). While addressing the urgent need to reduce environmental impacts the solution continues to support the prevalent unsustainable systems of production and consumption. The macro-scale challenge of altering these systems remains. Sustainable transition theories are not revolutionary, calling for the overthrow or transformation of capitalist neoliberal societies. My study, and the MLP theory I have adopted, can then be criticized for focusing on the meso level and failing to offer solutions at the macro level. Transitioning to renewable carbon materials alone will not solve the overarching problem of unsustainable consumption.

However, recycling plastic is an essential requirement of developing a circular economy, which many scholars agree is preferable to the prevailing linear system. Dramatically increasing the use of recycled plastics will act as a contestation, stimulating debate around consumption and spotlighting the benefits of a circular economy. Increasing the use of renewable feedstocks displaces demand for non-renewable fossil fuels, delivering an environmental win in the short term while creating an opportunity for the existing regime to reconfigure and reduce carbon impacts. Developing design projects using renewable plastics helps focus attention on the long-term (un)sustainability challenges facing the entrenched petrochemical regime and boost demand for these materials.

In 1977, Penny Sparke reviewed Jeffrey Meikle's book on plastics claiming the essential argument of his text is an 'inherent ambiguity within plastics as it simultaneously expresses our optimism and pleasure at being able to conquer nature … and at the same time our fear that this Faustian interference will end in disaster' (Sparke, 1997, p. 110). Without radical change the prophecy will come true; belief in the mastery of nature through technology increasingly challenged by a growing realization of the ecological repercussions of our choices. With the global population forecast to reach over 10 billion by the end of the century, the demand for chairs and most other consumer products will continue to grow. Constantly changing cultural, social, economic and technological conditions will ensure a continuous stream of new designs of chairs and all other consumer products, whether the world really needs them or not. Products made from plastics are often the least bad alternative to meet these demands. Designers, materials manufacturers, product makers and end-users must all accept their responsibility to preference renewable plastics wherever possible.

This book is a call to arms for urgent action to mitigate the most devastating impact of the climate emergency. Design can and must take a central role in delivering solutions to environmental and social problems. The way that products and services are designed will play a deterministic role in the success of our response to the climate emergency. With 80 per cent or more of a product's environmental impact determined during the design stage responsibility falls heavily on industrial designers (Resnick, 2019, p. 26). This study has shown that the agency of the designer is often limited; even internationally acclaimed designers are most often engaged on a project basis, with future employment dependent on client satisfaction. The structure of those relationships tends to diminish the agency of the dispensable designer. Despite those limitations, designers are uniquely placed to showcase and drive cultural acceptance, or even preference, for the unique attributes of renewable plastics, promoting their use across consumer goods.

The most environmentally friendly plastic can be made using renewable energy from organically farmed stover or from abundant natural resources such as algae or CO_2, coloured and strengthened with benign bio-based pigments and additives. Unfortunately, such materials remain rare. However, viable alternatives to virgin fossil plastics are available, as demonstrated by the environmentally conscious designers and manufacturers already experimenting with these materials across a wide range of consumer goods. While it is often easier and more cost-effective to continue using familiar material and manufacturing technologies, it is becoming increasingly clear that fossil plastics are taking us along a pathway to self-destruction. Expanding the use of renewable plastics in consumer goods guides us to a more sustainable, circular future, where materials are either reused or biodegrade at the end of a product's useful life.

With the industry producing hundreds of millions of tonnes of plastic every year it is tempting to trivialize the actions of individual actors. Can modest initiatives achieve anything beyond alleviating individual culpability for inaction on climate change, especially given the industry's steadfast commitment to produce even more virgin plastic? Rapid expansion of plastic production is driven by the relentless pursuit of profits to offset the impact of declining demand for oil. Plastics are the last bastion for the besieged fossil industry. The regime's dependence on a niche market (plastics) to drive projected growth is also the fossil regime's greatest vulnerability. Reducing the compound annual growth rate (CAGR) for plastic demand to 1 per cent (from 4 per cent) will be sufficient to significantly disrupt the business plans of the petrochemical industry, jeopardizing profits for shareholders (Carbon Tracker Initiative, 2020, p. 21).

By betting on continued expansion, the regime has elevated design to a powerful position. Replacing a quarter of the plastic used to make consumer durables with renewable plastics will result in the desired decrease in forecast growth, irrespective of the actions being undertaken across packaging and other sectors. Design should be a futuring agent of change and, particularly in the case of plastics, design can play a transformative role in shaping a more sustainable and forward-looking future.

Appendix

1. ERPR tool

The Environmentally Responsible Product Rating (ERPR) tool described in Chapter 4 is calculated by summing the scores awarded for five attributes: materials, weight, transport, appeal and end-of life prospects. This iteration of the tool has been developed specifically to enable a comparison of the environmental impacts of chairs made from renewable plastics using only the information typically available from manufacturers. Details of how scores were awarded for each attribute are given below. It is important to note that scores have been awarded based on claims made by manufacturers and have not been independently verified.

a. Materials

Material selection for a plastic chair (or any plastic product that does not consume resources during use) is the most important decision facing a designer or manufacturer, as the resources consumed producing plastic represents the largest proportion of a product's environmental impact. Renewable plastics can demonstrate favourable GHG impacts compared with traditional fossil-based materials. Bioplastic production emits about 80 per cent less carbon dioxide and consumes up to 65 per cent less energy during manufacture. In addition, bioplastics emit about 70 per cent less GHGs during degradation in landfill (Selvamurugan Muthusamy & Pramasivam, 2019). Recycled plastics deliver net saving of about 0.6 t of CO_2 per ton of plastic recycled compared with landfilling (WRAP UK, 2010).

As shown in Table 10.1 top scores of four points were awarded to monobloc designs made from a single material that is either produced from renewable organic sources or has been recycled in its entirety (over 90 per cent). By relying on a single material, monoblocs offer the advantage of eliminating the need to

Table 10.1 ERPR scoring sheet for materials

SCORE	MATERIAL
4 (n=33)	90%+ single material (recycled or organic)
3 (n=10)	90%+ recycled/organic materials (mixed e.g. legs different material)
2 (n=3)	75%+ recycled/organic materials mixed together
1 (n=13)	>50%,<75% recycled/organic materials mixed together OR legs not made from recycled material
0 (n=1)	<50% recycled/organic content

n=number of chair designs allocated to each category

join elements, thereby avoiding the use of screws, adhesives and welding. Where designs rely on two fully recycled or organic materials a score of three points is awarded, as separate manufacturing processes and assembly are required.

Where different materials are mixed or the legs of a chair have been made from virgin material, then the design is marked down. Some designs feature recyclates that have been mixed with other materials, to create compounds. Although mixing materials can reduce the end-of-life prospects for a product, this factor is not considered here, as it is evaluated separately. Only the content of the materials is evaluated when awarding the materials score.

b. Weight

After 'avoid', 'reduce' is the preferred strategy according to the waste hierarchy. Reducing the amount of material consumed delivers environmental savings throughout the life cycle of a product, including reduced transport and end-of-life impacts. Weight provides a good indication of the embodied energy contained within a chair. In the absence of detailed information on the production processes involved with creating each chair (and the sources of energy used), weight has been used as a proxy indicator of the resources consumed in the manufacturing process. It is acknowledged that weight might be a poor indicator of the total resources consumed during production but, without access to details about the source of each component and the resources consumed during their production and transportation, a more detailed calculation cannot be made. The lower the weight of the final product the higher the score, reflecting the assumed reduction in resource consumed during manufacture, with subsequent transportation and end-of-life impacts similarly reduced (Table 10.2).

Table 10.2 ERPR scoring sheet for weight

SCORE	KG
4 (n=6)	<3
3 (n=7)	3–3.99
2 (n=21)	4–4.99
1 (n=16)	5–6.99
0 (n=10)	7+

Note: Where more than one variant of the same design is available the weight of the lightest model has been rated except where a model is available with a base made from recycled or organic material, as these models score more highly for both material and end-of-life prospects.
n=number of chair designs allocated to each category

c. Transport

Most manufacturers produce chairs in a single location and then distribute globally. Without detailed information on the distribution of sales and methods of transport, it is not possible to accurately calculate the comparative efficiency of distribution strategies. However, the efficiency with which chairs can be transported is a function of their volume and weight. Designing chairs to be stacked during transportation reduces resource consumption, as volume is reduced, together with the quantity of packaging required. This is important throughout the supply chain, as the environmental impacts caused by shipping small quantities of chairs short distances (for example, from retail warehouses to customer premises) by diesel truck can significantly exceed the contribution to impacts made during a long journey on a container ship. Designing chairs for self-assembly is another approach to reducing transport impacts, as separate components can occupy less shipping space compared with fully assembled chairs.

Designs that can only be shipped singly in cartons represent the least efficient use of transport and have been awarded a score of zero points (Table 10.3). Scores

Table 10.3 ERPR scoring sheet for transport

SCORE	TRANSPORT
4 (n=6)	Stacked 6+
3 (n=15)	Stacked 4–5
2 (n=8)	Stacked or carton 2–3
1 (n=6)	3D printed
0 (n=25)	Shipped singly

Note: Stacked height rated is without the use of a trolley. Where chairs are shipped dismantled a bonus point has been awarded.

n=number of chair designs allocated to each category

awarded reflect the efficiency with which products are shipped which may differ from how efficiently products can be stacked while in use. For example, Emeco's *Broom Stacking Chair* can stack up to six units high, but is shipped in cartons of two units, therefore, a score of only two points was awarded. Chairs that were designed to be shipped dismantled were awarded a bonus point, to reflect the additional shipping volume reduction achieved. Where insufficient information is provided by manufacturers it is assumed products are shipped singly, fully assembled.

While knockdown and stacking are well-established solutions to reduce transport impacts, 3D printing promised to eliminate them. Although products can be produced at or near their point of consumption raw materials (filament) must still be transported to the printing location. Shipping relatively small quantities of raw materials to supply individual printing outlets potentially causes significantly greater impacts that those obtained by shipping large volumes of plastic to factories for injection-moulding chairs which are then distributed in bulk to retail warehouses. For these reasons, 3D printed chairs were awarded a score of just one point for transportation efficiency.

d. End-of-life

Most chairs are designed to last years of use, but furniture often gets discarded due to changing needs, taste or fashions as well as due to damage or wear and tear.

Many plastic chairs (including all monoblocs) cannot be disassembled or have parts replaced in the event of breakage. Plastic chairs often cannot be upgraded to alter their look, while chairs made with traditional materials can be re-polished or reupholstered to suit current trends. Additionally, plastic chairs are often used outdoors, which can significantly shorten their life expectancy. If colours and surfaces deteriorate plastic chairs become increasingly likely to be discarded. Consideration, therefore, needs to be given to what will happen to the chair at the end of its useful life.

The final destination for a plastic chair (or any other product) largely depends on where and how it is disposed of in real life. Even fully recyclable chairs can end up in landfill if they are disposed of inappropriately, especially in remote or low-income areas where recycling infrastructure is scarce or non-existent. When chairs are properly disposed of through the waste stream, they might be recycled, incinerated, used as fuel for energy from waste or end up in landfills, depending on local policies and infrastructure availability.

Some manufacturers acknowledge their producer responsibilities by offering take-back schemes to reuse or recycle their products. However, these schemes are often tokenistic, as the economic and environmental costs of returning unwanted furniture back to the manufacturer are prohibitive for most users.

While end-users have the ultimate responsibility for managing the end-of-life prospects for a chair, the designer and manufacturer can encourage circularity through: (limiting) their choice of materials, ensuring all plastic components are

Table 10.4 ERPR scoring sheet for end-of-life prospects

SCORE	END-OF-LIFE PROSPECTS
4 (n=10)	Fully recyclable (using existing infrastructure)
3 (n=10)	Fully recyclable (requires disassembly)
2 (n=6)	Recyclable/compostable (requires specialist facilities)
1 (n=11)	Recyclable through EPR scheme (or equivalent) OR partially recyclable (e.g. legs)
0 (n=23)	Not recyclable OR insufficient information available

n=number of chair designs allocated to each category

permanently labelled using the internationally recognized resin indication codes (RIC) and by designing for disassembly. Monoblocs made from a single material that can potentially be recycled using mechanical waste processing facilities, were awarded the top score of four points (Table 10.4). Designs that require manual disassembly, or materials that require specialist infrastructure to facilitate their sorting, recycling or composting, were awarded lower scores. Chairs made from bioplastics tended to perform poorly against this criterion, as many of these materials will only degrade when exposed to consistent heat and humidity conditions that only exist at industrial composting facilities, which remain scarce.

e. Appeal

This study presents a unique attempt to quantify a product's appeal and integrate it into an assessment environmental impact. Appeal is crucial for a product to succeed commercially, effectively delivering environmental benefits by displacing the purchase of a less-sustainable alternative.

I have considered three easily quantifiable criteria as crucial to creating broad appeal for of a chair: its price, the range of colours available and the number of variants or configurations available. Retail prices of the chairs included in this study (converted to Australian dollars from the retail price in the country of manufacture) were divided into quintiles to determine scores. Plastics possess a distinct personality which, as Karana et al. observed: 'lend themselves, particularly to brightly coloured, light-hearted and humorous, design' (Karana et al., 2013, p. xx). However, when using recyclates or bioplastics, designers may need to compromise on colour choices. Recycled plastics are often only available in a limited range of (often muted) colours. The lack of bright colours can make renewable plastics less appealing than their traditional fossil-based counterparts. Despite those limitations offering more colour choices is assumed to broaden the appeal of a design. In addition to colour, offering more variants to a basic design is likely to broaden the appeal of a chair. This approach enables designers to adapt chairs to meet the specific needs of consumers and businesses across a range of tasks. Monoblocs are disadvantaged when evaluating variants, they are usually only available in a single configuration.

The appeal score is calculated as the arithmetic mean of the three scores achieved for each dimension as shown in Table 10.5. A perfect score of four points indicates a design offered at a comparatively low price (<\$200) and available with a wide choice of colours (10+) and a variety of configurations (10+). It

Table 10.5 ERPR scoring sheet for appeal

SCORE	PRICE	COLOURS	VARIANTS
4 (n=0)	<$200 (12)	8+ (9)	10+ (2)
3 (n=2)	$200–$299 (12)	6–7 (11)	6–9 (2)
2 (n=29)	$300–$399 (12)	4–5 (10)	3–5 (2)
1 (n=19)	$400–$599 (10)	2–3 (20)	2 (14)
0 (n=10)	$600+ (14)	1 (10)	1 (40)

Note: Retail price in country of origin converted to AUD as at December 2022.
n=number of chair designs allocated to each category

is recognized that achieving a perfect score is extremely challenging. In practice designers and manufacturers must strike a balance between delivering variants and colour choices against offering the lowest possible retail price. Therefore, it is essential to consider these trade-offs when attempting to maximize a chair's appeal.

There are many other aspects to appeal, perhaps most importantly how the comfort offered by a chair is perceived. Additionally, some designs included in this analysis look very different from a typical chair, which might deter purchasers. Many renewable plastic chairs feature slightly rough or textured surfaces compared with virgin plastics which might deter some buyers, while others embrace the opportunity for a new sensual experience. Recycled plastics can display impurities in their surface, in contrast to the uniform finish of virgin plastics, which could reduce their appeal to both designers and purchasers. Such differences are challenging to quantify and their individual impact on the overall appeal and sales of a design is difficult to assess. To avoid these complexities, this analysis is restricted to evaluating criteria for which quantitative data is readily available.

Endnotes

Introduction

1. Michael Byrne, climate researcher at the University of Oxford contributor to the IPCC report.
2. Sotheby's London, November 2015, Lot 161.

Chapter 1

1. The economic advantage offered by synthetic plastics proved irresistible to producers as well – DuPont claimed a single operator could produce 10,000 combs in one day using an early (small) injection-moulding machine, compared with just 350 when working by hand (Freinkel, 2011b).
2. Writing in 1940, Van Doren appreciated the significance of plastics for 'the mass production designer. Here was a material that offered a chance for infinitely delicate detail; they were pleasant to the touch and lustrous, and offered an infinite colour range. Small wonder that designers welcomed them with open arms' (Van Doren, 1940, p. 303).
3. Brochure kindly supplied by Jeffrey Meikle.
4. The glass transition temperature is when the physical properties of plastics change from being rigid and solid to become more flexible or rubbery, different from the melting point when the substance begins to flow.
5. In the United States, 85 per cent of output from the plastics industry was diverted to the war effort (Young America Films, 1945), while in the UK 80 per cent of furniture manufacturers were involved with 'some sort of war work' (Kries et al., 2019, p. 354) In Australia, 'the Australian plastics industry, which concentrated almost 100% on war goods during 1939-45, is now sturdily pushing ahead with peacetime expansion' ('New Australian Plastics', 1946, p. 20).
6. See for example Bailey (2010) and 'The Bars Go Up' (1940). An article in *Time* magazine highlighted the impact on women: 'The U.S. woman, it appeared, by 1941s end would have a choice of 1) going barelegged, 2) buying Nylon stockings which might be unprocurable, or 3) wearing cotton stockings' ('The Presidency The Last Step Taken', 1941).
7. Production of plastics in the United States alone nearly quadrupled, from $213 million in 1939 to $818 million in 1945 (J. Meikle, 1995, p. 125). In Australia, *Pix* magazine reported: 'Well on its feet throughout the world before the war, the [plastics] industry benefited in wartime by finding new uses for its products, because of shortage of materials like rubber, metal' ('New Australian Plastics', 1946).
8. As early as 1940, Van Doren observed: 'In the proper use of plastics you will find a real challenge. Before their special beauties were fully understood, they were abominably

misused, to the point where bad design definitely retarded their advancement' (Van Doren, 1940, p. 303).

9. More mechanized production techniques using spray guns were introduced from the late 1950s (J. Meikle, 1995, p. 194).

10. However, *Tulip Chairs* failed to be accepted by the commercial market, the pedestal bases did not find favour with planners of office interiors (Lutz et al., 2012, p. 182).

11. A video showing the highly labour-intensive production process still required to make Yrjö Kukkapuro's Karuselli chair is available on Artek's website, www.artek.fi/en/products/karuselli-lounge-chair (accessed 5 March 2024).

12. Injection moulding was first developed in 1872 and advanced by German industrialists in the 1920s. James Hendry introduced screw injection moulding in 1946 (J. Meikle, 1995, pp. 80–84).

13. For example, in the furniture industry, Australian manufacture Aristoc was sold to Furniture Makers of Australia in the early 1970s (who had already purchased major competitor, Fler). Market conditions were similar in the UK where Kandya merged with Meredew, in 1968 (Isaac, 2017, p. 237).

14. Gerald Summers' *Bent Plywood Armchair* of 1933 is a notable exception.

15. In the 1920s, tubular steel and leather had been used by at least four German designers (Ludwig Mies van der Rohe, Marcel Breuer, Willem Gispen and Mart Stam) to develop cantilevered seating solutions, with other materials used for upholstery or to provide support. In the 1930s, Gerrit Rietveld developed the *Zig Zag* chair using solid wood, having previously experimented with plywood and steel to develop a continuous shape. This design also fell short of being a monobloc as it is constructed from four separate flat boards.

16. Panton studied architecture at the Royal Danish Academy of Fine Arts in Copenhagen before going on to work with Arne Jacobsen (designer of the *Swan, Egg* and *Drop* chairs, among others) from 1950 to 1952 where curved plywood was being used extensively (Engholm et al., 2018, p. 34).

17. Panton worked with another company (A. Sommer, distributed by Thonet) to develop a plywood version of this design (*S-chair*) which was available from 1965 or 1966 (Postell, 2012, p. 228).

18. Grosfillex's vice president of manufacturing explained that an expensive mould (well over $1 million) 'might be used to make a million copies of a monobloc for the Western market and after, five or seven years, the mould would be sold to a less demanding manufacturer in Africa for $50,000, and they will make another million chairs with the same mould and sell them very inexpensively' (Gosnell, 2004, p. 76).

19. Even labelling becomes redundant as makers' marks and logos can be stamped or embossed directly into the product during the moulding process.

20. Walter Benjamin claimed 'that which withers in the age of mechanical reproduction is the aura of the work of art' (Benjamin, 1935, p. 221).

Chapter 2

1. 'The potential market for modern plastic furniture is the 25–35-year age groups who are less conformist and show a decided swing away from traditional furniture' (Shaw, n.d., p. 31).

2. Several polls, conducted in 1970, showed that Americans quickly forgot the names of the astronauts who had been celebrated as heroes just one year before. For example, a *Philadelphia Sunday Bulletin* poll found 70 per cent of locals were unable to remember Neil Armstrong's name (Tribbe, 2014, p. 9).

3. Italian manufacturer Kartell was an early convert to celebrating the material in its own right – plastic-as-plastic, producing an ever-increasing range of brightly coloured consumer products with sleek surfaces. Science journalist Susan Freinkel observed: 'Kartell's designs made it possible for people to believe that plastic, like traditional materials had some noble presence' (Freinkel, 2011a, p. 38).

4. For example, in March 1988, a student paper reported that 'Since the publication of books like Rachel Carson's *Silent Spring* the idea of progress has never been quite the same. Too many people have become painfully aware that the triumphs of technology are often accompanied by "unintended consequences"' (Beder, 1986).

5. For a full discussion on the impacts of Earthrise see R. Poole (2010).

6. During 1973 plastic furniture featured on at least fifty pages of the two main Australian homemaker titles. By 1975, I counted only twenty pages, including modular furniture.

7. Agenda 21 of the United Nations Conference on Environment and Development, 1992.

8. *La Marie* was released at €250, considerably less than other offerings by named designers at Kartell at that time.

9. Air injection had previously been used to make parts of chairs (for example, Alberto Homann's *Ensemble Chair* for Fritz Hansen in 1992), but the *Air Chair* is claimed as the first to be made entirely using this process.

Chapter 3

1. According to an LCA analysis by CleanMetrics Corp.

2. Wrapping a cucumber in plastic can extend its shelf life from three days to two weeks (Dhall et al., 2012). A comprehensive European survey of fifty-seven products packaged in a range of plastics and alternative materials reported energy savings of 50 per cent by using plastic packaging. For details of these and other similar studies see Andrady and Andrady (2015, pp. 126, 129).

3. PPE includes face masks, ventilators, sterile gloves, eye visors, hazmat suits, N95 masks, visors, shoe covers and goggles, all mostly made from PP.

4. Broadcast in the UK from October 2017 and in Australia from February 2018.

5. 'Defuturing' is the negation of world futures for us, and many of our unknowing non-human others (Fry, 2020, p. 10).

6. China now produces 30 per cent of the world's plastic, doubling their market share over the past decade (European Environment Agency, 2021, p. 19).

7. The Nova Institute is a private research institute established to promote the transition of the chemical and material industry to renewable carbon.

8. Chemical Recycling Europe defines chemical recycling as 'any reprocessing technology that directly affects either the formulation of the polymeric waste or the polymer itself and converts them into chemical substances and/or products whether for the original or other purposes, excluding energy recovery'.

9. For a detailed review of CO_2 capture and use technologies see Pires da Mata Costa et al. (2021).
10. For a summary of how war in the Middle East has affected oil prices since 1973 see Ruiz Estrada et al. (2020, p. 3).
11. Although this includes 'credits', details of which were not supplied. Plans to use Virtual Power Purchase Agreements are mentioned (DuPont, 2021, p. 33).
12. First-generation bioplastics are made from food crops, subsequent generations can be made from non-food crops or stover (second-generation), algae or bacteria (third generation – although some definitions also include stover in this category) and CO_2 (fourth generation). An extensive review of relevant LCA studies (from 2020) found that higher generation feedstocks solve the problems of land use and food competition associated with first- and second-generation feedstocks. This reduces emissions during cultivation; however, higher input intensity (especially of energy) is currently required in the refinery stages, but this may be a reflection of the embryonic nature of these materials (Wellenreuther & Wolf, 2020).
13. The global bioplastics market size was valued at $10.2 billion in 2021 and is expected to achieve a compound annual growth rate (CAGR) of 17.1 per cent from 2022 to 2030 (Grand View Research, 2022a).
14. Despite these commitments, demand for virgin plastic from signatories increased from 11.8 million metric tonnes in 2021 to 11.9 million tonnes in 2022 (Ellen MacArthur Foundation, 2023, p. 23).

Chapter 4

1. Although it should be noted that these studies overlook the significant ocean bound microplastics created during the washing and shredding processes involved in mechanical recycling and the prevalence of harmful chemicals found in recycled plastics both of which are only recently been investigated (Brown et al., 2023; Carmona et al 2023).
2. Where more than one variant of a design is available the lightest chair in the series has been evaluated, except where a version is available with legs made from recycled or organic material, as these models score better for materials and end-of-life prospects. Other variants of the same design might have different environmental impacts. Some chairs are available with optional pads or cushions made from various materials designed to enhance comfort, but these have not been evaluated and may alter the environmental credentials of a design.
3. Grcic and Magis deliberately limited colour choice to three options for the *Bell Chair*. In an interview for this study Grcic explained that this decision minimized the wastage caused by changing colour, allowing production runs to be optimized to contain costs.

Chapter 5

1. Sustain-ability (in contrast to sustainability) acknowledges the need for change, based on the growth of the 'development of ecological sustainment'. Such 'development' implies the adoption of fundamental directional change, the creation of new economies and a rejection of a 'steady state' model as the other of existing forms of growth (Fry, 2009, p. 46).

2. A gloss meter measures specular reflection, the ratios of incident light and reflected light, by measuring the amount of light reflected from a beam applied to the surface allowing makers to precisely control the visual appearance of plastic products (Plastics Technology, 2020).

3. Emeco have since addressed this issue by working with the material supplier (BASF) to improve the polymer, which can now be injected at lower temperatures, producing a more consistent finish, nearly equivalent to the virgin product.

4. More recent versions of the Butter series are made from 100 per cent post-consumer plastic died to a range of colours.

Chapter 6

1. Only a few studies have adopted a prospective lens regarding the future upscaling potential of niche innovations (van Waes et al., 2018, p. 1300).

2. Throughout the 1940s oil averaged $1.55 per barrel, less than the $1.65 average for the 1920s (US Energy Information Administration, 2022).

3. For a detailed analysis of the impact of China Sword on the United States recycling industry see Heiges & O'Neill (2022).

4. In March 2020, more than two years after the China Sword policy came into effect, the Council of Australian Governments produced a white paper on how to respond to the need to develop a local recycling industry. When the promised AU$190 million funding for new waste infrastructure is made available, interested parties will need to participate in a competitive grants process. Successful applicants will then need to apply for planning permission (likely to be opposed by NIMBY action groups), arrange long-term supply commitments for feedstocks from local councils and build and commission the new recycling infrastructure, a process that is likely to take several years. Much of this funding is expected to support the processing of mixed waste, plastic and tyres into Refuse-Derived Fuels. In February 2021 (three years after China Sword), the NSW government announced its first grants of up to AU$5 million (AU$35 million in total), to support the waste industry in its response to Australia's waste export bans.

5. The eco-label index lists 456 eco-labels in use across 199 countries, https://www.ecolabelindex.com/ (accessed 9 March 2024).

6. The Center for International Environmental Law defines Cancer Alley as an industrial chemical corridor along the Mississippi River, an eighty-five mile stretch between New Orleans and Baton Rouge, Louisiana. The stretch contains seven out of ten census tracts with America's highest cancer rates.

7. For example, the Northern Territory government operates the longest running CDS in Australia and claims that 'The proportion of regulated containers in the litter stream decreased from five to ten percent prior to the commencement of the CDS to 3.1% in the first year of the commencement of the CDS, and maintained an average of 3.1% across the first five years of operation of the CDS' (Department of Environment and Natural Resources, 2018).

8. Directive (EU) 2018/852 of the European Parliament and of the Council of 30 May 2018 amending Directive 94/62/EC on packaging and packaging waste.

9. In 2020/1 AMCO claims only 18 per cent of plastic packaging was recycled or composted in Australia.
10. An estimated 129 billion face masks were being used every month during the height of the pandemic. In May 2021, Braskem reported the shortage of demand for durables (cars, electronics, etc.) was counter-balanced with a peak in demand for medical and hygiene products, as well as for packaging products (Xu & Ren, 2021). During 2020, Amazon generated an estimated 599 million pounds of plastic packaging waste, representing a 29 per cent increase on the previous year (Oceana, 2021, p. 4).

Chapter 7

1. At the start of 2023, A Plastic Planet launched *PlasticFree*, a materials database aimed as supporting designers working in product/packaging, textiles and the built environment seeking to work with more sustainable materials. See https://aplasticplanet.com (accessed 9 March 2024).
2. In particular, nitrous oxide, used in cheap fertilizers and favoured for non-food crops, can have significant GHG impacts.
3. Architect William McDonough and chemist Michael Braungart introduced the term 'cradle-to-cradle', in their book of the same name, which helped to popularize the circular economy concept. Their highly influential work specifically excluded plastics as they recognized that recycling invariably leads to downcycling, resulting in materials of a lesser quality. Despite the popularity of this work the myth that plastics can be continuously recycled endures (McDonough & Braungart, 2002).
4. For an overview of these recycling technologies see CSIRO (2021b).
5. For a regularly updated list of current and planned chemical recycling initiatives see www.sustainableplastics.com/news/chemical-recycling-tracker (accessed 9 March 2024).
6. Conversely, as at April 2024, at least twenty-five states have reclassified chemical recycling as manufacturing, exposing them to less stringent air and water restrictions.
6. In 2022, 2,149 patents for plastic recycling were filed, an eightfold increase since 2016 (Mathys & Squire, 2022).
7. For example, a 2020 LCA-based study claimed that 'it does not seem feasible to replace all the petrochemical plastic packaging with bioplastic because this will inevitably result in a considerable increase of land and water use' (Brizga et al., 2020, p. 45). This study concluded that satisfying global demand for packaging with bioplastics 'would require a minimum 61 million ha of land (which is larger than the total area of France) and at least 388.8 billion m3 of water (sixty per cent more than the EU's annual freshwater withdrawal)' (Brizga et al., 2020, p. 50). In contrast Zheng and Suh found, in 2019, that a complete shift of 250 million tonnes plastics production to bio-based plastics would only require 5 per cent of all arable land (Zheng & Suh, 2019, p. 374).
8. Including thermal (gasification), chemical catalytic (synthesis or dehydration) or biochemical (fermentation) (Bennett, 2012a, pp. 319–357).
9. 'Compostable' plastics should degrade into humus with an absence of toxic chemicals in line with international standards. At least 90 per cent of the compostable polymers

must be converted to carbon dioxide in industrial composting plants within six months, with remaining particles below 2 mm diameter (Gironi & Piemonte, 2011).

10. For a list of academic references detailing the health impacts of PFAS see Greenpeace International (2020, p. 16). For a study showing that PFAS migrate from food packaging see Schwartz-Narbonne et al. (2023). For a detailed analysis of the societal costs associated with PFAS see ChemSec (2023).

11. In a 2006 standard (GB/T 20197-2006).

12. *Güthoff v Deutsche Umwelthilfe* (2014).

13. *Ellepot v Sungrow* (2019).

14. During the summer of 2018, the European BioForever project commissioned sixty in-depth interviews with consumers, aged twenty-five to fifty-five, across Poland, Germany and Italy. Fossil plastics were seen as non-degradable by consumers while products based on plants were considered degradable. Not all bioplastics are degradable and some fossil plastics do degrade. Participants did not fully understand the difference between 'degradable' and 'biodegradable' (Carus et al., 2019).

15. For example, PET, the most commonly recycled plastic can be contaminated if mixed with just 0.1 per cent of PLA, reducing the transparency of recycled plastic. At 0.3 per cent the recycled PET would become opaque and show yellowing. If the PLA content is increased to 2–5 per cent, the different melting temperature could result in clusters of PLA forming, hampering recycling efforts (Greenpeace International, 2020, p. 24).

16. For more on the challenges of processing waste bioplastics see Bio-Plastics Europe (2023).

17. This was not a new idea, in the 1940s, Chicago architects Brenner, Speyer and Prestini entered a lounge chair made from resin impregnated wood fibre to the MoMA competition (Nelson, 1994, p. 57).

18. Household mixed plastic waste contains approximately 32 MJ per kg, which is favourable compared to coal (approximately 25 MJ per kg). Hence, plastics can substitute fossil fuels in heat or power production which is preferable to landfill but less preferable than recycling (Bennett, 2012a, p. 342).

19. As at the end of 2020, 250 major FMCG businesses committed to achieve four key commitments by 2025: eliminating all single-use plastics packaging which is problematic or unnecessary; ensuring that all plastic packaging is reusable, recyclable or compostable; ensuring that 50 per cent of plastic packaging is effectively recycled or composted and reaching 30 per cent recycled or responsibly sourced, bio-based content in packaging (Ellen MacArthur Foundation, 2020, p. 5).

Chapter 8

1. In June 2021, Audi announced they had taken the mass-balanced approach a step further, successfully producing car parts from recycled mixed plastics using a similar process (with a different material manufacturer) but claim that parts 'could' be manufactured entirely from pyrolysis oil. Closing the material cycle in this way, it is argued, can save valuable resources, energy and reduce GHG emissions. Although details of the energy required to collect and recycle the plastic waste together with the GHG impacts of consuming the pyrolysis oil to produced new plastics are not given (Audi, 2021).

2. DuPont have also joined the Plastics Europe Mass Balance Taskforce to investigate the potential of this approach (DuPont, 2021, p. 29).
3. www.ikea.com/au/en (accessed 12 September 2023).
4. Prices on the Australian IKEA website as at December 2021, September 2022 and September 2023.
5. For a review of the eight voluntary eco-labelling schemes in use by the furniture industry in the EU plus six more from surrounding countries see Broeck (2015). For a more focused discussion of the confusion this causes see Donatello et al. (2020).
6. Caution should be exercised here, as studies have shown consumers are confused by and lack trust in eco-labels (Joshi & Rahman, 2015, p. 138).
7. For very high volumes injection moulding is likely to remain the preferred technology. However, in an increasingly fragmented market it is likely that even large manufacturers will be incentivized to experiment with technologies more suited to smaller production runs.
8. For example, Hartmut Esslinger (who has created many successful designs for Apple and Sony among other) claims 'for the power is in the factories – and beyond that, however old they are, designers need to know what factories can do today. Because the best designs and products always emerge through collaboration with the manufacturing side' (Bürdek, 2005, p. 92).
9. In Grcic's case the engineer spoke Italian while Grcic's first language is German.
10. The only exception to this is Emeco, which has specialized in using recycled materials since the 1940s.
11. Designfabrik offers design and materials advice to designers, engineers and developers in the early stages of product development (designfabrik, n.d.). The division was launched in 2006 in an apparent attempt at revisiting the strategy of knowledge transfer successfully employed by the German plastics industry up until the 1970s (Streb, 2003).
12. The bio version of the *Componibili* is available from Kartell in the UK (as at October 2022), although priced at a 46 per cent premium to the standard version.
13. Having been offered for auction three times, starting with expectations of over €100 million, a potential sale was announced in September 2022 for just €17 million, less than 4 per cent of the value implied by the Kartell purchase (Barrett, 2022b).

Conclusion

1. Based on the average 4 per cent of annual plastic production that ends up in our oceans and the estimated 129 billion facemasks being disposed of every month (Xu & Ren, 2021).
2. See: www.climateemergencydeclaration.org/climate-emergency-declarations-cover-15-million-citizens (accessed 9 March 2024).
3. However, as Escobar observes, transitioning to a post-extractivist framework does not 'endorse a view of untouched nature, nor a ban on all mining or larger-scale agriculture, but rather the significant transformation of these activities so as to minimize their environmental and cultural impact'. The future for plastics is unlikely to be idealistic but consist of a patchwork of solutions, with the incumbent petrochemical regime adapting to take a significant role as pioneered by BASF and Braskem.

4. BASF has multiple projects to reduce emissions including electrification, renewable energy and plastic recycling, among others (Pires da Mata Costa et al., 2021). Evonik offer a version of nylon12, made using renewable energy in Germany and reducing GHG emissions by 40 per cent.

5. This is especially relevant following the launch of 'Cracker of the Future', a consortium including BASF, Borealis, BP, LyondellBasell, SABIC and Total aiming to electrify steam crackers, replacing the use of fossil fuels by renewable electricity (Duckett, 2021).

6. Since interviews were completed for this study Plastic Free have launched a subscription-based service giving access to a growing database of plastic-free alternative materials. https://plasticfree.com (accessed 9 March 2024).

Bibliography

A good image through good design. (1971). *Plastics News*, April.

Aarnio, E., & Savolainen, J. (2016). *Eero Aarnio: Designer of colour and joy* (A. Svenskberg, ed.). WSOY.

Abdelmoez, W., Dahab, I., Ragab, E. M., Abdelsalam, O. A., & Mustafa, A. (2021). Bio- and oxo-degradable plastics: Insights on facts and challenges. *Polymers for Advanced Technologies*, 32(5), 1981–1996. https://doi.org/10.1002/pat.5253.

Akhurst, S. (2004). The rise and fall of Melamine tableware. *The Plastiquarian*, 32, 8–13.

All plastic home on way. (1935, August 12). *North Western Courier*, 5.

Alliance to End Plastic Waste. (2022). *Feedstock quality guidelines for pyrolysis of plastic waste*. https://endplasticwaste.org/en/our-stories/feedstock-for-pyrolysis (accessed 9 March 2024.

Altherr, J. (2020, April). Soft shapes for soft lounge—furniture and design latest. https://www.arper.com/ww/en/magazine/brand-stories/soft-shapes-for-soft-lounge (accessed 9 March 2024).

Amadei, A. M., Sanyé-Mengual, E., & Sala, S. (2022). Modeling the EU plastic footprint: Exploring data sources and littering potential. *Resources, Conservation and Recycling*, 178, 106086. https://doi.org/10.1016/j.resconrec.2021.106086.

Ambasz, E. (1972). *Italy: The new domestic landscape achievements and problems of Italian design*. MoMA.

American Chemistry Council. (2019). Population growth and materials demand study (PLASTICS). https://www.americanchemistry.com/better-policy-regulation/plastics/resources/population-growth-and-materials-demand-study (accessed 9 March 2024).

Andrady, A. L., & Andrady, A. L. (2015). *Plastics and environmental sustainability*. John Wiley & Sons, Inc.

Andrews, D. (2015). The circular economy, design thinking and education for sustainability. *Local Economy*, 30(3), 305–315. https://doi.org/10.1177/0269094215578226.

Anssary, A. E. (2006). *An approach to support the design process considering technological possibilities* PhD, Universidad de Duisburg-Essen.

APCO. (2019). Australia's 2025 national packaging targets. https://apco.org.au/national-packaging-targets (accessed 9 March 2024).

Ashby, M. F., & Johnson, K. (2003). The art of materials selection. *Materials Today*, 6(12), 24–35. https://doi.org/10.1016/S1369-7021(03)01223-9.

Ashby, M. F., & Johnson, K. (2014). *Materials and design: The art and science of material selection in product design*. Elsevier Science & Technology.

Attfield, J. (2000). *Wild things: The material culture of everyday life*. Berg.

Audi. (2021, June 17). A new lease on life: Recycling automotive plastics, Audi MediaCenter. https://www.audi-mediacenter.com/en/press-releases/a-new-lease-on-life-recycling-automotive-plastics-14024 (accessed 9 March 2024).

Audibert, K. (2010). *65–75: Jean-Pierre Laporte, dix ans de création*. K. Audibert.

Australian Plastics. (1945, 22 June). *Illawarra Mercury*, 2.

Back to Blue. (2023). Peak plastics: Bending the consumption curve. https://backtoblueinitiative.com/plastics-consumption/ (accessed 9 March 2024).

Bacon, F. (1942, 27 October). Science in the home. *Woman's Mirror*, 18(49). https://nla.gov.au/nla.obj-505817600.

Bailey, R. H. (2010, August). Iron will. *World War II; Leesburg*, 25(2), 50–57.

Barrett, A. (2019a, 9 June). Thai government gives tax deduction for using bioplastics packaging. *Bioplastics News*. https://bioplasticsnews.com/2019/06/09/thai-government-gives-tax-deduction-for-using-bioplastics-packaging (accessed 9 March 2024).

Barrett, A. (2019b, 2 September). Best bioplastic company of 2019 Bio-on. *Bioplastics News*. https://bioplasticsnews.com/2019/09/03/best-bioplastic-company-2019-award (accessed 9 March 2024).

Barrett, A. (2022a, 31 August). INEOS and Covestro partner on mass balanced polycarbonate. *Bioplastics News*. https://bioplasticsnews.com/2022/08/31/ineos-and-covestro-partner-on-mass-balanced-polycarbonate (accessed 9 March 2024).

Barrett, A. (2022b, 19 September). Maip to buy Bio-on (FREE). *Bioplastics News*. https://bioplasticsnews.com/2022/09/19/maip-to-buy-bio-on-free (accessed 9 March 2024).

Bartlett, N. (1946, 28 November). The plastic refrigerator is here (Wot, no beer?). *Daily Telegraph*, 23.

Basel Action Network. (2024). *Australia export data 2023*. https://www.ban.org/plastic-waste-project-hub/trade-data/australia-export-data-annual-summary (accessed 9 March 2024).

BASF. (2022). Chemical recycling of plastic waste. https://www.basf.com/global/en/who-we-are/sustainability/we-drive-sustainable-solutions/circular-economy/mass-balance-approach/chemcycling.html (accessed 9 March 2024).

Beder, S. (1986, 18 March). The social shaping of technology. *Tharunka*, 29.

Benjamin, W. (1935). The work of art in the age of mechanical reproduction. In *Illuminations* (p. 26). Harcourt, Brace & World.

Bennett, S. J. (2012a). Implications of climate change for the petrochemical industry: Mitigation measures and feedstock transitions. In W.-Y. Chen, J. Seiner, T. Suzuki, & M. Lackner (Eds.), *Handbook of climate change mitigation* (pp. 319–357). Springer US. https://doi.org/10.1007/978-1-4419-7991-9_10.

Bennett, S. J. (2012b). Using past transitions to inform scenarios for the future of renewable raw materials in the UK. *Energy Policy*, 50, 95–108. https://doi.org/10.1016/j.enpol.2012.03.073.

Bensaude-Vincent, B. (2013). Plastics, materials and dreams of dematerialization. In *Accumulation: The material politics of plastic* (pp. 17–29). Routledge, Taylor & Francis Group.

Berkhout, F., Smith, A., & Stirling, A. (2004). Socio-technological regimes and transition contexts. In *System innovation and the transition to sustainability* (pp. 48–75). Edward Elgar Publishing.

Beyond Plastics. (2021). *The new coal: Plastics & climate change*. Bennington College. https://www.beyondplastics.org/plastics-and-climate (accessed 9 March 2024).

Bickersteth, R. (2024, 7 March). Vitra 'willing to take risks' over changing products' appearance to improve sustainability, says CEO Nora Fehlbaum. Dezeen. https://www.dezeen.com/2024/03/07/vitra-sustainabiity-ceo-nora-fehlbaum-interview (accessed 9 March 2024).

Biessey, P., Vogel, J., Seitz, M., & Quicker, P. (2023). Plastic waste utilization via chemical recycling: Approaches, limitations, and the challenges ahead. *Chemie Ingenieur Technik*, 95(8), 1199–1214. https://doi.org/10.1002/cite.202300042.

Bio-on. (2017, December 20). Kartell enters Bio-on to help develop new technologies. GlobeNewswire News Room. https://www.globenewswire.com/news-release/2017/12/20/1266757/0/en/Kartell-enters-Bio-on-to-help-develop-new-technologies.html (accessed 9 March 2024).

Bio-Plastics Europe. (2023). *Handbook on the impacts of bio-based and biodegradable plastics on existing waste management frameworks*. https://bioplasticseurope.eu/downloads/public-deliverables (accessed 9 March 2024).

Bioplastics Magazine. (2020, 2 January). Bio-on officially declared bankrupt. https://www.bioplasticsmagazine.com/en/news/meldungen/20200101Bio-on-officially-declared-bankrupt.php (accessed 9 March 2024).

Bjørn, A., & Hauschild, M. Z. (2013). Absolute versus relative environmental sustainability. *Journal of Industrial Ecology*, 17(2), 321–332. https://doi.org/10.1111/j.1530-9290.2012.00520.x.

Blum, A. (2008, 18 August). Q&A: Philippe Starck on bioplastics, Virgin Galactic, and his impossible chair. *Wired*. https://www.wired.com/2008/08/pl-design-9 (accessed 9 March 2024).

Boehnert, J. (2018). *Design, ecology, politics: Towards the ecocene*. Bloomsbury Academic.

Boeing. (2014). *Boeing: 2014 environment report*. http://www.exi.boeing.com/aboutus/environment/environment_report_14/index.html (accessed 9 February 2022).

Bony, A., & Rakocevic, I. (Eds.). (2005). *Furniture & interiors of the 1970s*. Flammarion.

Boyd, J. (2011). Celluloid: The eternal substitute. *Science History Institute*. https://www.sciencehistory.org/distillations/celluloid-the-eternal-substitute (accessed 9 March 2024).

Braskem. (2018, 20 July20 July). Braskem's green plastic now in chairs released by Tramontina. https://www.braskem.com.br/news-detail/braskems-green-plastic-now-in-chairs-released-by-tramontina (accessed 9 March 2024).

Braskem. (2022, 25 October). Braskem invests in capacity expansion and partnerships for the production of biobased plastics. https://www.braskem.com.br/europe/news-detail/braskem-invests-in-capacity-expansion-and-partnerships-for-the-production-of-biobased-plastics (accessed 9 March 2024).

Break Free From Plastic. (2021). Missing the mark: Unveiling corporate false solutions to the plastic crisis. https://www.breakfreefromplastic.org/missing-the-mark-unveiling-corporate-false-solutions-to-the-plastic-crisis (accessed 9 March 2024).

Bremmer, J. (2022). Australian refuse derived fuel: Fuel product of plastic waste export in disguise? National Toxics Network/IPEN.

Brizga, J., Hubacek, K., & Feng, K. (2020). The unintended side effects of bioplastics: Carbon, land, and water footprints. *One Earth*, 3(1), 45–53. https://doi.org/10.1016/j.oneear.2020.06.016.

Broeck, E. V. D. (2015, 14 April). The EU furniture market situation and a possible furniture products initiative. *CEPS*. https://www.ceps.eu/ceps-publications/eu-furniture-market-situation-and-possible-furniture-products-initiative (accessed 9 March 2024).

Brown, E., MacDonald, A., Allen, S., & Allen, D. (2023). The potential for a plastic recycling facility to release microplastic pollution and possible filtration remediation effectiveness. *Journal of Hazardous Materials Advances*, 10, 100309. https://doi.org/10.1016/j.hazadv.2023.100309.

Bruggers, J. (2019, 25 February). Plastics: The new coal in Appalachia? Inside Climate News. https://insideclimatenews.org/news/25022019/plastics-hub-appalachian-fracking-ethane-cracker-climate-change-health-ohio-river (accessed 9 March 2024).

Buchholz, K., Beil, R., Institut Mathildenhöhe, & Ausstellung (Eds.). (2007). *Plexiglas, Werkstoff in Architektur und Design*. Wienand.

Bui, S., Cardona, A., Lamine, C., & Cerf, M. (2016). Sustainability transitions: Insights on processes of niche-regime interaction and regime reconfiguration in agri-food systems. *Journal of Rural Studies*, 48, 92–103. https://doi.org/10.1016/j.jrurstud.2016.10.003.

Burall, P. (1996). *Product development and the environment*. Gower.

Buranyi, S. (2018, 13 November). The plastic backlash: What's behind our sudden rage – and will it make a difference? *The Guardian*. https://www.theguardian.com/environment/2018/nov/13/the-plastic-backlash-whats-behind-our-sudden-rage-and-will-it-make-a-difference (accessed 9 March 2024).

Bürdek, B. E. (2005). *Design history, Theory and practice of product design*. de Gruyter.

Byars, M., & d'Altoé, C. A. (2006). *New chairs: Design, technology, and materials*. Laurence King.

Carbon Tracker Initiative. (2020). The future's not in plastics: Why plastics demand won't rescue the oil sector. Carbon Tracker Initiative. https://carbontracker.org/reports/the-futures-not-in-plastics/ (accessed 9 March 2024).

Carlsen, N. (2019, 12 March). TangForm. http://tangform.com (accessed 11 February 2022).

Carmona, E., Rojo-Nieto, E., Rummel, C. D., Krauss, M., Syberg, K., Ramos, T. M., Brosche, S., Backhaus, T., & Almroth, B. C. (2023). A dataset of organic pollutants identified and quantified in recycled polyethylene pellets. *Data in Brief*, 51, 109740. https://doi.org/10.1016/j.dib.2023.109740.

Carpenter, E., & Smith, K. L. (1972). Plastics on the Sargasso Sea surface. *Science*, 175(4027), 1240–1241.

Carter, L. (2021, 1 July). *Inside Exxon's playbook*. Unearthed. https://unearthed.greenpeace.org/2021/07/01/exxon-undercover-pfas-plastic-chemicals/ (accessed 11 June 2024).

Carus, M., Partanen, A., & Piotrowski, S. (2019). *Bio-based products from forestry via economically viable European routes*. Nova Institute.

Cary, C. M., DeLoid, G. M., Yang, Z., Bitounis, D., Polunas, M., Goedken, M. J., Buckley, B., Cheatham, B., Stapleton, P. A., & Demokritou, P. (2023). Ingested polystyrene nanospheres translocate to placenta and fetal tissues in pregnant rats: Potential health implications. *Nanomaterials*, 13(4), Article 4. https://doi.org/10.3390/nano13040720.

Center for International Environmental Law. (2019). *Oil, gas and the climate: An analysis of oil and gas industry plans for expansion and compatibility with global emission limits.* https://www.ciel.org/reports/oil-gas-and-climate-an-analysis-of-oil-and-gas-industry-plans-for-expansion-and-compatibility-with-global-emission-limits/ (accessed 9 March 2024).

Centre for Climate Integrity. (2024). The fraud of plastic recycling. https://climateintegrity.org/plastics-fraud (accessed 9 March 2024).

Chang, J. (2016). *Commentary: China coal-to-olefins (CTO) investment to slow.* ICIS Explore. https://www.icis.com/explore/resources/news/2016/05/26/10002356/commentary-china-coal-to-olefins-cto-investment-to-slow (accessed 11 June 2024).

Changing Markets Foundation. (2020). Talking trash. Changing Markets Foundation. https://talking-trash.com (accessed 9 March 2024).

Changwichan, K., Silalertruksa, T., & Gheewala, S. H. (2018). Eco-efficiency assessment of bioplastics production systems and end-of-life options. *Sustainability*, 10(4), 952. https://doi.org/10.3390/su10040952.

Chapman, W. (2020, May 28). Demand will drive soft plastics recycling. http://sustainabilitymatters.net.au/content/waste/article/demand-will-drive-soft-plastics-recycling-1311372333 (accessed 9 March 2024).

Chaudhuri, S., & Eaton, C. (2020, 15 October). Firms like Dow bet billions on plastics. Now there's a glut. *Wall Street Journal*. https://www.wsj.com/articles/firms-like-dow-bet-billions-on-plastics-now-theres-a-glut-11602754200 (accessed 9 March 2024).

Chemical Guys. (n.d.). Chemical guys new car smell air freshener & odor eliminator. https://www.chemicalguys.com/new-car-scent/new-car-scent.html (accessed 6 October 2022).

ChemSec. (2023). The top 12 PFAS producers in the world and the staggering societal costs of PFAS pollution. https://chemsec.org/reports/the-top-12-pfas-producers-in-the-world-and-the-staggering-societal-costs-of-pfas-pollution (accessed 9 March 2024).

CIEL and Break Free from Plastics. (2022). Winter is coming: Plastic has to go. https://www.breakfreefromplastic.org/winter-is-coming (accessed 9 March 2024).

Clean Air Council. (2022, 2 February). Environmental groups take legal action against PA chemical plant for air pollution violations. https://cleanair.org/environmental-groups-take-legal-action-against-pa-chemical-plant-for-air-pollution-violations (accessed 9 March 2024.

Cleminshaw, D. (Ed.). (1989). *Design in plastics: Successful product design in plastics.* Rockport Publishers.

Coca, N. (2020, 23 January). Asian countries spurn and burn waste imports. China Dialogue. https://chinadialogue.net/en/cities/11801-asian-countries-spurn-and-burn-waste-imports (accessed 9 March 2024).

Collins, J. B. (1988). 'Design in Industry' exhibition, National Gallery of Canada, 1946: Turning Bombers into lounge chair. *Material Culture Review/Revue de La Culture Matérielle*, 27, 27–38.

Cranz, G. (1998). *The chair: Rethinking culture, body, and design* (1st ed.). Norton.

Creadore, L. T., & Castaldi, M. J. (2022). Quantitative comparison of life cycle assessments of advanced recycling technologies for end-of-life plastics. *Journal of Energy Resources Technology*, 145(042201). https://doi.org/10.1115/1.4055843.

Crystal-Like Furniture. (1940, 8 January). *Mercury (Hobart)*, 10.

CSIRO. (2021a). A circular economy roadmap for plastics, tyres, glass and paper in Australia. CSIRO. https://www.csiro.au/en/research/natural-environment/circular-economy (accessed 9 March 2024).

CSIRO. (2021b). Advanced recycling. https://research.csiro.au/ending-plastic-waste/advanced-recycling (accessed 9 March 2024).

Dalberg. (2019). No plastic in nature: Assessing plastic ingestion from nature to people. University of Newcastle. https://www.wwf.org.au/news/news/2019/revealed-plastic-ingestion-by-people-could-be-equating-to-a-credit-card-a-week (accessed 9 March 2024).

Davidson, D. J., Jones, K. E., & Parkins, J. R. (2016). Food safety risks, disruptive events and alternative beef production: A case study of agricultural transition in Alberta. *Agriculture and Human Values*, 33(2), 359–371. https://doi.org/10.1007/s10460-015-9609-8.

Day, K. (2011, 28 March). An interview with Dirk Vander Kooij. COOL HUNTING®. https://coolhunting.com/design/dirk (accessed 9 March 2024).

De Oliveira, F. C., & Coelho, S. T. (2017). History, evolution, and environmental impact of biodiesel in Brazil: A review. *Renewable and Sustainable Energy Reviews*, 75, 168–179. https://doi.org/10.1016/j.rser.2016.10.060.

Dennis, L. (2020). *A matter of Material: Exploring the value of the Museum of Design in Plastics*. PhD, University of Brighton.

Department of Environment and Natural Resources. (2018). *Evaluation of the operation of the Northern Territory container deposit scheme*. Northern Territory Government. https://ntepa.nt.gov.au/publications-and-advice/container-deposit-scheme-reports (accessed 9 March 2024).

Design for Modern Living. (1940, April). *The DuPont Magazine*, 35(4), 7–9.

Design plays a stellar role in exports. (1980). *Australian Export Furniture Journal*.

Designboom. (2018, 5 May). Jasper Morisson's 1 inch reclaimed chair for emeco. *Designboom*. https://www.designboom.com/design/jasper-morisson-1-inch-reclaimed-chair-emeco-05-05-2018 (accessed 9 March 2024).

designfabrik. (n.d.). *Designfabrik®*. https://plastics-rubber.basf.com/global/de/performance_polymers/creation_center/designfabrik.html (accessed 5 October 2022).

Despeisse, M., Baumers, M., Brown, P., Charnley, F., Ford, S. J., Garmulewicz, A., Knowles, S., Minshall, T. H. W., Mortara, L., Reed-Tsochas, F. P., & Rowley, J. (2017). Unlocking value for a circular economy through 3D printing: A research agenda. *Technological Forecasting and Social Change*, 115, 75–84. https://doi.org/10.1016/j.techfore.2016.09.021.

Dewberry, E., & Johnson, J. (2010). *Design interventions, prediction and science in the sustainable transition of large complex systems*, 7.

Dhall, R. K., Sharma, S. R., & Mahajan, B. V. C. (2012). Effect of shrink wrap packaging for maintaining quality of cucumber during storage. *Journal of Food Science and Technology*, 49(4), 495–499. https://doi.org/10.1007/s13197-011-0284-5.

Diamantopoulos, A., Schlegelmilch, B. B., Sinkovics, R. R., & Bohlen, G. M. (2003). Can socio-demographics still play a role in profiling green consumers? A review of the evidence and an empirical investigation. *Journal of Business Research*, 56(6), 465–480. https://doi.org/10.1016/S0148-2963(01)00241-7.

Diaz, M., Darnhofer, I., Darrot, C., & Beuret, J.-E. (2013). Green tides in Brittany: What can we learn about niche-regime interactions? *Environmental Innovation and Societal Transitions*, 8, 62–75. https://doi.org/10.1016/j.eist.2013.04.002.

DiNoto, A., & Arky, D. (Eds.). (1984). *Art plastic: Designed for living* (2. Print). Abbeville Press.

Dodds & Shute. (2020). *Dodds & Shute sustainability report*. Dodds & Shute.

Donatello, S., Cordella, M., Kaps, R., & Kowalska, M. (2020). Are the existing EU ecolabel criteria for furniture products too complex? An analysis of complexity from a material and a supply chain perspective and suggestions for ways ahead. *The International Journal of Life Cycle Assessment*, 25, 868–882.

DuBois, J. H. (1972). *Plastics history U.S.A.* Cahners Books.

Duckett, A. (2021, 24 March). Greener chemicals: Steam cracking could go electric by 2023. *The Chemical Engineer*. https://www.thechemicalengineer.com/news/greener-chemicals-steam-cracking-could-go-electric-by-2023 (accessed 9 March 2024).

DuPont. (2021). *Thriving together (sustainability report)*. DuPont de Nemours Inc. https://www.dupont.com/about/sustainability-2021.html (accessed 5 October 2022).

Egenhoefer, R. B. (2019). *Routledge handbook of sustainable product design*. Routledge, Taylor & Francis Group.

Ehrlich, P. R. (1968). *The population bomb*. Buccaneer Books.

Eisner, K. R. (1976, 27 January). PVC food containers. *Canberra Times*, 8.

El Bilali, H. (2019). The multi-level perspective in research on sustainability transitions in agriculture and food systems: A systematic review. *Agriculture*, 9(4), Article 4. https://doi.org/10.3390/agriculture9040074.

Ellen MacArthur Foundation. (2016). The new plastics economy: Rethinking the future of plastics. World Economic Forum, Ellen MacArthur Foundation and McKinsey & Company. https://ellenmacarthurfoundation.org/the-new-plastics-economy-rethinking-the-future-of-plastics (accessed 9 March 2024).

Ellen MacArthur Foundation. (2017). New plastics economy—the future of plastics. https://www.newplasticseconomy.org/ (accessed 5 October 2022).

Ellen MacArthur Foundation. (2020). *The global commitment 2020*. https://www.ellenmacarthurfoundation.org/resources/apply/global-commitment-progress-report (accessed 5 October 2022).

Ellen MacArthur Foundation. (2022). About us: What we do. https://ellenmacarthurfoundation.org/about-us/what-we-do (accessed 5 October 2022).

Ellen MacArthur Foundation. (2023). *The global commitment 2023*. https://www.ellenmacarthurfoundation.org/global-commitment-2023/overview (accessed 9 March 2024).

Elzen, B. (2004). *System innovation and the transition to sustainability theory, evidence and policy*. Edward Eldgar.

Engholm, I., Michelsen, A., & Panton, V. (2018). *Verner Panton: Environments, colours, systems, patterns*. Phaidon.

Environmental Integrity Project. (2024*).* Billions in taxpayer subsidies to U.S. plastics plants support illegal air pollution in communities of color. https://environmentalintegrity.org/news/billions-in-taxpayer-subsidies-to-u-s-plastics-plants-support-illegal-air-pollution/ (accessed 11 June 2024).

Escobar, A. (2018). *Designs for the pluriverse: Radical interdependence, autonomy, and the making of worlds.* Duke University Press.

European Bioplastics. (2020, 2 December). Market update 2020. *European Bioplastics e.V.* https://www.european-bioplastics.org/market-update-2020-bioplastics-continue-to-become-mainstream-as-the-global-bioplastics-market-is-set-to-grow-by-36-percent-over-the-next-5-years (accessed 9 March 2024).

European Environment Agency. (2021). Plastics, a growing environmental and climate concern. News. https://www.eea.europa.eu/highlights/plastics-environmental-concern (accessed 9 March 2024).

European Investment Bank. (2023). Cutting plastic pollution. https://www.eib.org/en/press/all/2023-084-plastic-pollution-new-study-finds-at-least-6-7-billion-investment-gap-to-meet-europe-plastics-recycling-targets (accessed 11 June 2024).

Fachagentur Nachwachsende Rohstoffe e. V. (2016). *Processing of bioplastics.* http://ifbb-knvb.wp.hs-hannover.de/db (accessed 5 October 2022).

Fallan, K. (2019). *The culture of nature in the history of design.* Taylor & Francis Group.

Featherston, G. (1971). The future of plastics in furniture. *Australian Furniture Trades Journal.*

Ferguson, S. M. (2012). *Plastics without petroleum history and politics of 'Green' plastics in the United States.* PhD, Rensselaer Polytechnic Institute.

Fiell, C., & Fiell, P. (2009). *Plastic dreams: Synthetic visions in design.* Fiell Publ.

Fiell, C., Fiell, P., & Muthesius, A. (1993). *Modern chairs.* Benedikt Taschen.

Fisher, T. (2013). A world of colour and bright shining surfaces: Experiences of plastics after the Second World War. *Journal of Design History*, 26(3), 285–303. https://doi.org/10.1093/jdh/ept012.

Fisher, T. (2015). Fashioning plastics. In *The Social Life of Materials* (pp. 119–135). Routledge, Taylor & Francis Group.

Fisher, T. H. (2004). What we touch, touches us: Materials, affects, and affordances. *Design Issues*, 20(4), 20–31. https://doi.org/10.1162/0747936042312066.

Fleck, H. R. (1944). *Whither plastics?* Temple Press.

Fler increase their use of plastics. (1973). *Plastics News*, May.

Food & Water Watch. (2019). *The fracking endgame: Locked into plastics, pollution, and climate chaos.* https://www.foodandwaterwatch.org/wp-content/uploads/2021/03/rpt_1905_fracking-2019-web_2.pdf (accessed 9 March 2024).

Fortune Business Insights. (2023). Plastics market size, share | Global industry forecast [2030]. https://www.fortunebusinessinsights.com/plastics-market-102176 (accessed 9 March 2024).

Francis, R. (2016). *Recycling of polymers: Methods, characterization and applications.* John Wiley & Sons, Inc.

Frazier, R. (2022a, 6 October). An ethane cracker in western Pa. will soon start up. *StateImpact Pennsylvania.* https://stateimpact.npr.org/pennsylvania/2022/10/06/an-ethane-cracker-in-western-pa-will-soon-start-up-we-answered-your-questions-about-it/ (accessed 9 March 2024).

Frazier, R. (2022b, 15 November). Shell's ethane cracker, a mammoth plastics plant near Pittsburgh, begins operations. *StateImpact Pennsylvania.* https://stateimpact.npr. org/pennsylvania/2022/11/15/shells-ethane-cracker-a-mammoth-plastics-plant-near-pittsburgh-begins-operations (accessed 9 March 2024).

Frearson, A. (2023, 15 August). 'Egg chair would not be designed today' say Luke Pearson and Tom Lloyd. Dezeen. https://www.dezeen.com/2023/08/15/egg-chair-criticism-recycled-luke-pearson-tom-lloyd (accessed 9 March 2024).

Freinkel, S. (2011a). *Plastic: A toxic love story.* Text Publishing.

Freinkel, S. (2011b, 29 May). A brief history of plastic's conquest of the world. *Scientific American.* https://www.scientificamerican.com/article/a-brief-history-of-plastic-world-conquest (accessed 9 March 2024).

Friedel, R. D. (1983). *Pioneer plastic: The making and selling of celluloid.* University of Wisconsin Press.

Friedrichs, A., & Eickhoff, H. (Eds.). (2010). *220°C virus monobloc: The infamous chair.* Gestalten.

'Frightening' risk in foam stuffing. (1982, 8 July). *Canberra Times,* 6.

Fry, T. (1999). *A new design philosophy: An introduction to defuturing.* UNSW Press.

Fry, T. (2009). *Design futuring: Sustainability, ethics, and new practice* (English ed.). Berg.

Fry, T. (2020). *Defuturing: A new design philosophy.* Bloomsbury Visual Arts.

Fry, T., & Nocek, A. (2021). *Design in crisis.* Taylor & Francis Group.

Fuenfschilling, L., & Truffer, B. (2014). The structuration of socio-technical regimes—Conceptual foundations from institutional theory. *Research Policy,* 43(4), 772–791. https://doi.org/10.1016/j.respol.2013.10.010.

Fuller, R. B. (1969). *Operating manual for spaceship earth.* Southern Illinois University Press.

Gabrys, J., Hawkins, G., & Michael, M. (Eds.). (2013). *Accumulation: The material politics of plastic.* Routledge, Taylor & Francis Group.

GAIA. (2022, August). Legislative alert | Tracking trends in advanced/chemical 'recycling'. GAIA. https://no-burn.mystagingwebsite.com/resources/advanced-recycling-legislative-alert (accessed 9 March 2024).

Gaziulusoy, A. (2010). *System innovation for sustainability: A scenario method and a workshop process for product development teams.* PhD, University of Auckland.

Geels, F. (2007). Typology of sociotechnical transition pathways. *Research Policy,* 36(3), 399–417. https://doi.org/10.1016/j.respol.2007.01.003.

Geels, F., & Deuten, J. J. (2006). Local and global dynamics in technological development: A socio-cognitive perspective on knowledge flows and lessons from reinforced concrete. *Science & Public Policy (SPP),* 33(4), 265–275. https://doi.org/10.3152/147154306781778984.

Geels, F. W. (2002). Technological transitions as evolutionary reconfiguration processes: A multi-level perspective and a case-study. *Research Policy,* 31(8), 1257–1274. https://doi.org/10.1016/S0048-7333(02)00062-8.

Geels, F. W. (2005a). Processes and patterns in transitions and system innovations: Refining the co-evolutionary multi-level perspective. *Technological Forecasting and Social Change*, 72(6), 681–696. https://doi.org/10.1016/j.techfore.2004.08.014.

Geels, F. W. (2005b). *Technological transitions and system innovations: A co-evolutionary and socio-technical analysis*. Edward Elgar Publishing.

Geels, F. W. (2010). Ontologies, socio-technical transitions (to sustainability), and the multi-level perspective. *Research Policy*, 39(4), 495–510. https://doi.org/10.1016/j.respol.2010.01.022.

Geels, F. W. (2011). The multi-level perspective on sustainability transitions: Responses to seven criticisms. *Environmental Innovation and Societal Transitions*, 1(1), 24–40. https://doi.org/10.1016/j.eist.2011.02.002.

Geels, F. W., & Schot, J. (2007). Typology of sociotechnical transition pathways. *Research Policy*, 36(3), 399–417. https://doi.org/10.1016/j.respol.2007.01.003.

Geels, F. W., & Turnheim, B. (2022). *The great reconfiguration: A socio-technical analysis of low-carbon transitions in UK electricity, heat, and mobility systems*. Cambridge University Press. https://doi.org/10.1017/9781009198233.

Genus, A., & Coles, A.-M. (2008). Rethinking the multi-level perspective of technological transitions. *Research Policy*, 37(9), 1436–1445. https://doi.org/10.1016/j.respol.2008.05.006.

Gerassimidou, S., Lanska, P., Hahladakis, J. N., Lovat, E., Vanzetto, S., Geueke, B., Groh, K. J., Muncke, J., Maffini, M., Martin, O. V., & Iacovidou, E. (2022). Unpacking the complexity of the PET drink bottles value chain: A chemicals perspective. *Journal of Hazardous Materials*, 430, 128410. https://doi.org/10.1016/j.jhazmat.2022.128410.

Geyer, R., Jambeck, J. R., & Law, K. L. (2017a). Production, use, and fate of all plastics ever made. *Science Advances*, 3(7), e1700782. https://doi.org/10.1126/sciadv.1700782.

Geyer, R., Jambeck, J. R., & Law, K. L. (2017b). Production, use, and fate of all plastics ever made—Supplementary material. *Science Advances*, 3(7), e1700782. https://doi.org/10.1126/sciadv.1700782.

Gironi, F., & Piemonte, V. (2011). Bioplastics and petroleum-based plastics: Strengths and weaknesses. *Energy Sources, Part A: Recovery, Utilization, and Environmental Effects*, 33(21), 1949–1959. https://doi.org/10.1080/15567030903436830.

Gloag, J. (1945). *Plastics and industrial design*. George, Allen & Unwin Ltd.

Gosnell, M. (2004, July). Everybody take a seat. *Smithsonian; Washington*, 55(4), 74, 76–78.

Graedel, T. E. (1998). *Streamlined life-cycle assessment*. Prentice Hall.

Grand View Research. (2022a). *Bioplastics market size, share, growth report, 2020–2027*. https://www.grandviewresearch.com/industry-analysis/bioplastics-industry (accessed 5 October 2022).

Grand View Research. (2022b). *Furniture market size, share, growth & trends report, 2030*. https://www.grandviewresearch.com/industry-analysis/furniture-market (accessed 9 March 2024).

Grant, N. (2017). Mediating matters. In *Routledge handbook of sustainable product design* (pp. 222–235). Routledge, Taylor & Francis Group.

Greenpeace. (2017). Estimating carbon emissions from China's coal-to-chemical industry during the '13th five-year plan' period. https://www.greenpeace.org/eastasia/publication/984/estimating-carbon-emissions-from-chinas-coal-to-chemical-industry-during-the-13th-five-year-plan-period/ (accessed 11 June 2024).

Greenpeace International. (2020). Biodegradables will not solve China's plastics crisis. https://www.greenpeace.org/international/press-release/46066/biodegradables-will-not-solve-chinas-plastics-crisis (accessed 9 March 2024).

Guidot, R. (Ed.). (2006). *Industrial design techniques and materials*. Flammarion.

Hahn, J. (2023, 6 April). 'Formidable' Italian furniture brand Zanotta acquired by Cassina. *Dezeen*. https://www.dezeen.com/2023/04/06/zanotta-cassina-acquisition-haworth-lifestyle-design (accessed 9 March 2024).

Hansen, F. (2017). *In perfect shape*. teNeues Publishing UK Ltd.

Heffernan, M. (2022, 28 September). PureCycle opposition part of wider anti-plastic push. *Plastics Recycling Update*. https://resource-recycling.com/plastics/2022/09/28/purecycle-opposition-part-of-wider-anti-plastic-push/ (accessed 9 March 2024).

Heiges, J., & O'Neill, K. (2022). A recycling reckoning: How Operation National Sword catalyzed a transition in the U.S. plastics recycling system. *Journal of Cleaner Production*, 378, 134367. https://doi.org/10.1016/j.jclepro.2022.134367.

Höjer, M., & Mattsson, L.-G. (2000). Determinism and backcasting in future studies. *Futures*, 32(7), 613–634. https://doi.org/10.1016/S0016-3287(00)00012-4.

Holtz, G., Brugnach, M., & Pahl-Wostl, C. (2008). Specifying 'regime'—A framework for defining and describing regimes in transition research. *Technological Forecasting & Social Change*, 75(5), 623–643. https://doi.org/10.1016/j.techfore.2007.02.010.

Hopewell, J., Dvorak, R., & Kosior, E. (2009). Plastics recycling: Challenges and opportunities. *Philosophical Transactions of the Royal Society B: Biological Sciences*, 364(1526), 2115–2126. https://doi.org/10.1098/rstb.2008.0311.

Horani, L. F. (2020). Identification of target customers for sustainable design. *Journal of Cleaner Production*, 274, 123102. https://doi.org/10.1016/j.jclepro.2020.123102.

Howard, N. (2017). *The grail methodology for attributional life cycle assessment*.

HSBC Global Research. (2014). *Coal-to-chemicals*.

Hu, C. J., Garcia, M. A., Nihart, A., Liu, R., Yin, L., Adolphi, N., Gallego, D. F., Kang, H., Campen, M. J., & Yu, X. (2024). Microplastic presence in dog and human testis and its potential association with sperm count and weights of testis and epididymis. *Toxicological Sciences*, kfae060. https://doi.org/10.1093/toxsci/kfae060 (accessed 11 June 2024).

Huda, Z., & Bulpett, R. (2012). *Materials science and design for engineers*. Trans Tech Publications.

Hundertmark, T., Mayer, M., McNally, C., Simons, T. J., & Witte, C. (2019). How plastics waste recycling could transform the chemical industry. *Hydrocarbon Processing*, 9–13.

IKEA. (2020). *People & planet positive IKEA Group sustainability strategy for 2020*.

IKEA Australia. (n.d.). *Odger chair*. IKEA. https://www.ikea.com/au/en/p/odger-chair-anthracite-30457314 (accessed 9 March 2024).

International Energy Agency. (2018). The future of petrochemicals – Analysis. https://www.iea.org/reports/the-future-of-petrochemicals (accessed 9 March 2024).

International Energy Agency. (2020). *The oil and gas industry in energy transitions*. International Energy Agency.

IPSOS. (2019). A throwaway world: The challenge of plastic packaging and waste. https://www.ipsos.com/en/throwaway-world-challenge-plastic-packaging-and-waste (accessed 9 March 2024).

Irwin, T., & Kossoff, G. (2017). *2017 transition design seminar syllabus*. Carnegie Mellon University.

Isaac, G. (2017). *Featherston*. Thames & Hudson.

Jackson, L. (2011). *Robin & Lucienne Day: Pioneers of contemporary design*. Octopus.

Jambeck, J. R., Geyer, R., Wilcox, C., Siegler, T. R., Perryman, M., Andrady, A., Narayan, R., & Law, K. L. (2015). Plastic waste inputs from land into the ocean. *Science*, 347(6223), 768–771. https://doi.org/10.1126/science.1260352.

Jeong, H. J., & Ko, Y. (2020). Analysing the structure of bioplastic knowledge networks in the automotive industry. *International Journal of Technology Management*, 82(2), 132–150.

Johnson, E., & Vadenbo, C. (2020). Modelling variation in petroleum products' refining footprints. *Sustainability*, 12(22). https://doi.org/10.3390/su12229316.

Joore, J. P. (2010). *New to improve – The mutual influence between new products and societal change processes*. PhD, Delft University of Technology.

Jordans, F. (2022, 7 July). German lawmakers back plan to expand renewable energy. AP NEWS. https://apnews.com/article/russia-ukraine-germany-moscow-berlin-54f68326b9516d43dbbe2cde82c64dda (accessed 9 March 2024).

Joshi, Y., & Rahman, Z. (2015). Factors affecting green purchase behaviour and future research directions. *International Strategic Management Review*, 3(1), 128–143. https://doi.org/10.1016/j.ism.2015.04.001.

Karana, E. (2010). How do materials obtain their meanings? *METU Journal of the Faculty of Architecture*, 27(2), 271–285. https://doi.org/10.4305/METU.JFA.2010.2.15.

Karana, E., Elisa, G., & Valentina, R. (2017). Materially yours. In *Routledge handbook of sustainable product design* (pp. 206–221). Routledge.

Karana, E., Hekkert, P., & Kandachar, P. (2008). Material considerations in product design: A survey on crucial material aspects used by product designers. *Materials & Design*, 29(6), 1081–1089. https://doi.org/10.1016/j.matdes.2007.06.002

Karana, E., Pedgley, O., & Rognoli, V. (2013). *Materials experience: Fundamentals of materials and design*. Elsevier Science & Technology.

Katz, S. (1978). *Plastics: Designs and materials*. Studio Vista.

Katz, S. (1985). *Classic plastics: From Bakelite to high-tech: With a collector's guide*. Thames & Hudson.

Kemp, R. (1994). Technology and the transition to environmental sustainability: The problem of technological regime shifts. *Futures*, 26(10), 1023–1046. https://doi.org/10.1016/0016-3287(94)90071-X.

Kemp, R. P. M., Rip, A., & Schot, J. (2001). Constructing transition paths through the management of niches. In *Path dependence and creation* (pp. 269–299). Psychology Press. https://doi-org.ezproxy.lib.uts.edu.au/10.4324/9781410600370.

Kemp, R., & Rip, A. (1988). Technological change. In *Human choice and climate change* (vol. 1, pp. 327–399). Battelle Press.

Köhler, J., Geels, F. W., Kern, F., Markard, J., Wieczorek, A., Alkemade, F., Avelino, F., Bergek, A., Boons, F., Fünfschilling, L., Hess, D., Holtz, G., Hyysalo, S., Jenkins, K., Kivimaa, P., Martiskainen, M., McMeekin, A., Mühlemeier, M. S., Nykvist, B., … Wells, P. (2019). An agenda for sustainability transitions research: State of the art and future directions. *Environmental Innovation and Societal Transitions*, 31, 1–32. https://doi.org/10.1016/j.eist.2019.01.004.

Kopatz, V., Wen, K., Kovács, T., Keimowitz, A. S., Pichler, V., Widder, J., Vethaak, A. D., Hollóczki, O., & Kenner, L. (2023). Micro- and nanoplastics breach the blood–brain barrier (BBB): Biomolecular corona's role revealed. *Nanomaterials*, 13(8). https://doi.org/10.3390/nano13081404.

Körner, I., Redemann, K., & Stegmann, R. (2005). Behaviour of biodegradable plastics in composting facilities. *Waste Management*, 25(4), 409–415. https://doi.org/10.1016/j.wasman.2005.02.017.

Kries, M., Eisenbrand, J., Hoffmann, M., Agermann Ross, J., Gardner, C., Hale, C., Bassam, L., Koivu, A., & Vitra Design Museum (Eds.). (2022). *Plastic: Remaking our world* (1. Auflage). Vitra Design Museum.

Kries, M., Eisenbrand, J., & Saunders, B. (Eds.). (2019). *Atlas of furniture design*. Vitra Design Museum.

Laird, K. (2022, 7 January). Lack of trust deters use of recycled materials in high-end applications Bosch. *Sustainable Plastics*. https://www.sustainableplastics.com/news/lack-trust-deters-use-recycled-materials-high-end-applications (accessed 9 March 2024).

Laird, K. (2023, 9 March). EU plastics recycling capacity grows 17% in 2021. *Sustainable Plastics*. https://www.sustainableplastics.com/news/eu-plastics-recycling-capacity-grows-17-2021 (accessed 9 March 2024).

Lambert, S. (2021). *Provocative plastics: Their value in design and material culture*. Springer International Publishing AG.

Landon-Lane, M. (2018). Corporate social responsibility in marine plastic debris governance. *Marine Pollution Bulletin*, 127, 310–319. https://doi.org/10.1016/j.marpolbul.2017.11.054.

Landrigan, P., Raps, H., & Cropper, M. (2023). *Minderoo-Monaco commission on plastics and human health*. http://www.annalsofglobalhealth.org/collections/special/the-minderoo-monaco-commission-on-plastics-and-human-health (accessed 5 October 2022).

Latham, K. (2021, 12 May). The world's first 'infinite' plastic. https://www.bbc.com/future/article/20210510-how-to-recycle-any-plastic (accessed 9 March 2024).

Lauren, A. (2019, 31 October). Why Philippe Starck's ghost chair is here for eternity. Forbes. https://www.forbes.com/sites/amandalauren/2019/10/31/why-phillipe-starcks-ghost-chair-is-here-for-eternity (accessed 9 March 2024).

Lebreton, L., Slat, B., Ferrari, F., Sainte-Rose, B., Aitken, J., Marthouse, R., Hajbane, S., Cunsolo, S., Schwarz, A., Levivier, A., Noble, K., Debeljak, P., Maral, H., Schoeneich-Argent, R., Brambini, R., & Reisser, J. (2018). Evidence that the great Pacific garbage patch is rapidly accumulating plastic. *Scientific Reports*, 8(1), 1–15. https://doi.org/10.1038/s41598-018-22939-w.

Leschen, B. (1978). Plastics aired on talk-back radio. *Plastics News, Apil*.

Liboiron, M. (2013). Plasticizers: A twenty-first-century miasma. In *Accumulation: The material politics of plastic* (pp. 134–169). Routledge, Taylor & Francis Group.

Louisiana Bucket Brigade. (2020, 4 November). Army corps suspends permit for Formosa plastics' controversial Louisiana plant. *Louisiana Bucket Brigade*. https://labucketbrigade.org/army-corps-suspends-permit-for-formosa-plastics-controversial-louisiana-plant/ (accessed 9 March 2024).

Loultcheva, M. K., Proietto, M., Jilov, N., & La Mantia, F. P. (1997). Recycling of high density polyethylene containers. *Polymer Degradation and Stability*, 57(1), 77–81. https://doi.org/10.1016/S0141-3910(96)00230-3.

Lutz, B., Knoll, F., & Diffrient, N. (2012). *Eero Saarinen: Furniture for everyman*. Pointed Leaf Press.

Máčel, O., Woertman, S., & Wijk, C. van. (2008). *Chairs: The Delft collection*. 010 Publishers.

Mackenzie, D. (1997). *Green Design: Design for the environment* (2nd ed.). L. King.

Macrotrends. (n.d.). WTI crude oil prices—10 year daily chart. https://www.macrotrends.net/2516/wti-crude-oil-prices-10-year-daily-chart (accessed 5 October 2022).

Madden, O. (2012). Balancing ingenuity and responsibility in the age of plastic. *The age Of Plastic: Ingenuity and responsibility*, 2012 MCI Symposium, Suitland.

Madge, P. (1997). Ecological design: A new critique old. *Design Issues*, 13(2), 44–54. https://doi.org/10.2307/1511730.

Magnier, L., Mugge, R., & Schoormans, J. (2019). Turning ocean garbage into products – Consumers' evaluations of products made of recycled ocean plastic. *Journal of Cleaner Production*, 215, 84–98. https://doi.org/10.1016/j.jclepro.2018.12.246.

Manzini, E. (1992). Plastics and the challenge of quality. Unpublished. https://www.changedesign.org/Resources/Manzini/Manuscripts.htm (accessed 11 June 2024).

Manzini, E. (2009). New design knowledge. *Design Studies*, *30*(1), 4–12. https://doi.org/10.1016/j.destud.2008.10.001.

Manzini, E. (2015). Design in the transition phase: A new design culture for the emerging design. *Design Philosophy Papers*, 13(1), 57–62.

Manzini, E., & Cau, P. (1989). *The material of invention* (1st MIT ed.). MIT Press.

Manzini, E., Coad, R., Friedman, K., & Stolterman, E. (2015). *Design, when everybody designs: An introduction to design for social innovation*. MIT Press.

Marchese, K. (2020, 3 January). 'Recycling is only a bandaid,' says parley for the oceans founder Cyrill Gutsch. *Designboom*. https://www.designboom.com/design/ocean-plastic-cyrill-gutsch-interview-parley-for-the-oceans-03-01-2020 (accessed 9 March 2024).

Markard, J., Raven, R., & Truffer, B. (2012). Sustainability transitions: An emerging field of research and its prospects. *Research Policy*, 41(6), 955–967. https://doi.org/10.1016/j.respol.2012.02.013.

Marriott, J., & Minio-Paluello, M. (2013). Where does this stuff come from? Oil, plastic and the distribution of violence. In *Accumulation: The material politics of plastic* (pp. 171–183). Routledge, Taylor & Francis Group.

Martin, D. (1979). Furnishing trends for 1979. *The Australian Furnishing Trade Journal*.

Marusic, K. (2021, 10 February). Appalachia's fracking boom has done little for local economies: Study. EHN. https://www.ehn.org/fracking-economics-2650429410.html (accessed 9 March 2024).

Masson-Delmotte, V., Zhai, P., Pirani, S., & Connors, C. (2021). *IPCC, 2021: Climate change 2021: The physical science basis. Contribution of working group I to the sixth assessment report of the Intergovernmental Panel on Climate Change*. Intergovernmental Panel on Climate Change.

Mathys & Squire. (2022, 8 August). Patents for plastic recycling hit record high, up eightfold to 2,149 in the past five years. Mathys & Squire LLP. https://www.mathys-squire.com/insights-and-events/news/patents-for-plastic-recycling-hit-record-high-up-eightfold-to-2149-in-the-past-five-years (accessed 9 March 2024).

Matslinder. (2017, 21 January). Bendt Winge – Klassike of moderne mobler. *Mats Linder*. http://www.matslinder.no/2017/01/21/bendt-winge-klassiske-og-moderne-mobler (accessed 9 March 2024).

McDonald, H. (1973). Crisis and creativity in plastics furniture. *Plastics News*, May.

McDonough, W., & Braungart, M. (2002). *Cradle to cradle: Remaking the way we make things* (1st ed.). North Point Press.

McKinsey & Company. (2023). Boosting the supply of recycled materials for packaging. https://www.mckinsey.com/industries/paper-forest-products-and-packaging/our-insights/filling-the-gap-boosting-supply-of-recycled-materials-for-packaging (accessed 9 March 2024).

Meadows, D., Meadows, D., Randers, J., & Behrens, W. (1972). *The limits to growth*. Universe Books.

Meikle, J. (1995). *American plastic: A cultural history*. Rutgers University Press.

Mianehrow, H., & Abbasian, A. (2017). Energy monitoring of plastic injection molding process running with hydraulic injection molding machines. *Journal of Cleaner Production*, 148, 804–810. https://doi.org/10.1016/j.jclepro.2017.02.053.

Milliken. (2020, 11 December). Performance modifier additives make standard polypropylene resins perform like premium grades. https://www.ptonline.com/articles/game-changing-polypropylene-additives-improve-impact-resistance-and-processing-costs- (accessed 9 March 2024).

Milliken. (2021). At PRSE 2021 learn how Millikens additive solutions boost recycling and the circular economy. https://www.milliken.com/en-us/businesses/chemical/news/at-prse-2021-learn-how-millikens-additive-solutions-boost-recycling-and-the-circular-economy (accessed 9 March 2024).

Milne, R. (2023, 25 September). Lego ditches oil-free brick in sustainability setback. *Financial Times*. https://www.ft.com/content/6cad1883-f87a-471d-9688-c1a3c5a0b7dc (accessed 9 March 2024).

Modern Plastics. (1948, August). *Decoration and Glass*, 14(2), 52–54.

Moldenhauer, A. (2021, 29 January). Stefan Diez: Circular design guidelines. Stylepark. https://www.stylepark.com/en/news/stefan-diez-circular-design-guidelines-sustainability (accessed 9 March 2024).

Morello, A., & Ferrieri, A. (1988). *Plastic and design*. Arcadia.

Morozov, E. (2013). *To save everything, click here: Technology, solutionism and the urge to fix problems that don't exist*. Allen Lane.

NCP. (n.d.). *S-1500 Product sheet*. https://ncp.no/en/products/chairs/s-1500 (accessed 5 October 2022).

Nelson, G. (Ed.). (1994). *Chairs*. Acanthus Press.

New Australian Plastics. (1946, 20 April). *Pix*, 17(13), 20–22.

Nielsen, T. D., Hasselbalch, J., Holmberg, K., & Stripple, J. (2020). Politics and the plastic crisis: A review throughout the plastic life cycle. *WIREs Energy and Environment*, 9(1), e360. https://doi.org/10.1002/wene.360.

Niermann, I. (2004, April 26). Ingo Niermann: Plastic chair at functionalfate. org - Jens Thiel′s Monoblock Plastic Chairs Weblog. https://web.archive.org/web/20080121033144/http:/www.functionalfate.org/archives/2004/08/26/light-seems-heavy/ (accessed 9 March 2024).

Nova Institute. (2020a). Bio-based products: Green premium prices and consumer perception of different biomass feedstocks. https://renewable-carbon.eu/publications/product/bio-based-products-green-premium-prices-and-consumer-perception-of-different-biomass-feedstocks (accessed 9 March 2024).

Nova Institute. (2020b). The future of the chemical and plastics industry: Renewable carbon – bio-based news. http://news.bio-based.eu/the-future-of-the-chemical-and-plastics-industry-renewable-carbon/ (accessed 9 March 2024).

Noyes, E. (1941). *Organic design in home furnishings*. The Museum of Modern Art.

Oceana. (2021). Exposed: Amazon's enormous and rapidly growing plastic pollution problem. Oceana. https://oceana.org/reports/amazon-report-2021 (accessed 9 March 2024).

OECD. (2013). *Policies for bioplastics in the context of a bioeconomy* (OECD Science, Technology and Industry Policy Papers 10). https://doi.org/10.1787/5k3xpf9rrw6d-en.

Oladejo, J., & Rollinson, A. (2020). Chemical recycling: Status, sustainability, and environmental impacts. Global Alliance for Incinerator Alternatives. https://doi.org/10.46556/ONLS4535.

Paben, J. (2022, 28 September). Energy crisis threatens European plastics reclaimers. *Plastics Recycling Update*. https://resource-recycling.com/plastics/2022/09/28/energy-crisis-threatens-european-plastics-reclaimers (accessed 9 March 2024).

Paben, J. (2023, 28 August). Plastic reclaimers grapple with 'rock bottom' pricing. *Resource Recycling News*. https://resource-recycling.com/recycling/2023/08/28/plastic-reclaimers-grapple-with-rock-bottom-pricing (accessed 9 March 2024).

Panton Chair—Verner Panton—Official. (2018, 24 January). https://www.verner-panton.com/en/collection/panton-chair (accessed 9 March 2024).

Papanek, V. J. (1995). *The green imperative: Natural design for the real world*. Thames & Hudson.

Papanek, V. (2019). *Design for the real world* (2nd ed.). Thames & Hudson.

Pasca, V. (1991). *Vico Magistretti: Elegance and innovation in postwar Italian design*. Thames & Hudson.

Penna, C. C. R., & Geels, F. W. (2015). Climate change and the slow reorientation of the American car industry (1979–2012): An application and extension of the Dialectic Issue LifeCycle (DILC) model. *Research Policy*, 44(5), 1029–1048. https://doi.org/10.1016/j.respol.2014.11.010.

Pickering, G. (1975). Plastics in an energy short world. *Plastics News*, November.

Piercy, N. F. (2002). *Market-led strategic change: A guide to transforming the process of going to market* (3rd ed.). Butterworth-Heinemann.

Pires da Mata Costa, L., Micheline Vaz de Miranda, D., Couto de Oliveira, A. C., Falcon, L., Stella Silva Pimenta, M., Guilherme Bessa, I., Juarez Wouters, S., Andrade, M. H. S., & Pinto, J. C. (2021). Capture and reuse of carbon dioxide (CO2) for a plastics circular economy: A review. *Processes*, 9(5), Article 5. https://doi.org/10.3390/pr9050759.

PlastChem. (2024). State of the science on plastic chemicals. https://www.plasticpollutioncoalition.org/resource-library/plastchem-state-of-the-science-on-plastic-chemicals (accessed 11 June 2024).

Plastic Free July. (2021). *Impact report*. https://www.plasticfreejuly.org/blog/2021-annual-report (accessed 9 March 2024).

Plastics and you. (1976, 22 September). *Victor Harbour Times*, 7.

Plastics covered chair. (1951). *Australian Plastics*, September.

Plastics fumes 'hazard'. (1972, 19 June). *Canberra Times*, 3.

Plastics: Giant grows. (1956, 10 May). *Argus*, 13.

Plastics give luxury living. (1962, 31 January). *Biz*, 7.

Plastics in War. (1951). *Australian Plastics*, August.

Plastics Industry Association. (2023, 19 April). *Plastics survey: Consumers overwhelmingly support all types of recycling* [Text]. https://www.plasticsindustry.org/newsroom/plastics-survey-consumers-overwhelmingly-support-all-types-of-recycling (accessed 9 March 2024).

Plastics invade the home. (1952, 1 October). *Illawarra Daily Mercury*, 9.

Plastics made from starch. (1974). *Plastics News*, May.

Plastics Technology. (2020, 26 August). Quantifying the appearance of plastics. https://www.ptonline.com/articles/quantifying-the-appearance-of-plastics (accessed 9 March 2024).

Poole, J. (2020, 8 December). EUBP foresees dynamic global bioplastic growth inspired by portfolio diversification and COVID-19 navigation. https://www.packaginginsights.com/news/eubp-foresees-dynamic-global-bioplastic-growth-inspired-by-portfolio-diversification-and-covid-19-navigation.html (accessed 9 March 2024).

Poole, R. (2010). *Earthrise: How man first saw the earth*. Yale University Press.

Porter, B. (1972). Consumer attitudes to plastics housewares. *Plastics News*, August.

Postell, J. C. (2012). *Furniture design* (2nd ed.). John Wiley & Sons, Inc.

Recycling Markets. (2022, 14 September). Prices for scrap plastics continue to collapse. Plastics Recycling Update. https://resource-recycling.com/plastics/2022/09/14/prices-for-scrap-plastics-continue-to-collapse (accessed 9 March 2024).

Renewable Carbon Initiative. (2024, 23 February). Products made from crude oil have a significantly higher CO2 footprint than previously assumed. *Renewable Carbon News*. https://renewable-carbon.eu/news/products-made-from-crude-oil-have-a-significantly-higher-co2-footprint-than-previously-assumed (accessed 9 March 2024).

Resnick, E. (Ed.). (2019). *The social design reader* (1st ed.). Bloomsbury Visual Arts.

Reuters. (2019, 25 July). Italy's Bio-on shares hit hard on allegations of accounting flaws. *Reuters*. https://www.reuters.com/article/us-italy-bioon-idUSKCN1UK1M9 (accessed 9 March 2024).

Reuters. (2022, 2 March). Australia under fire for shipping plastic trash as 'fuel'. News. Trust.Org. https://news.trust.org/item/20220302114142-jskun/ (accessed 11 June 2024).

Rodriguez, A. (2016, 7 October). Siamese chair by Karim Rashid for A Lot of Brasil. 3rings. https://media.designerpages.com/2016/10/siamese-chair-by-karim-rashid-for-a-lot-of-brasil (accessed 9 March 2024).

Ross, J. (2020, 17 June). A Herculean task. *DisegnoDaily*. https://www.disegnodaily.com/article/a-herculean-task (accessed 5 October 2022).

Rubin, E. (2004). *Plastics and dictatorship in the German Democratic Republic: Towards an economic, consumer, design and cultural history*. PhD, The University of Wisconsin – Madison.

Ruiz Estrada, M. A., Park, D., Tahir, M., & Khan, A. (2020). Simulations of US-Iran war and its impact on global oil price behavior. *Borsa Istanbul Review*, 20(1), 1–12. https://doi.org/10.1016/j.bir.2019.11.002.

Ryan, C. (2009). Climate change and ecodesign, Part II: Exploring distributed systems. *Journal of Industrial Ecology*, 13(3), 350–353. https://doi.org/10.1111/j.1530-9290.2009.00133.x.

Rybczynski, W. (2017). *Now I sit me down: From klismos to plastic chair: A natural history*. Farrar, Straus and Giroux.

S-CAB. (n.d.). Chairs: Ginevra go green product sheet. https://www.scabdesign.com/en/products/chairs/ginevra-go-green-en.html (accessed 9 March 2024).

Schirmer, E., Schuster, S., & Machnik, P. (2021). Bisphenols exert detrimental effects on neuronal signaling in mature vertebrate brains. *Communications Biology*, 4(1), Article 465. https://doi.org/10.1038/s42003-021-01966-w.

Schot, J., & Geels, F. W. (2008). Strategic niche management and sustainable innovation journeys: Theory, findings, research agenda, and policy. *Technology Analysis & Strategic Management*, 20(5), 537–554. https://doi.org/10.1080/09537320802292651.

Schwartz-Narbonne, H., Xia, C., Shalin, A., Whitehead, H. D., Yang, D., Peaslee, G. F., Wang, Z., Wu, Y., Peng, H., Blum, A., Venier, M., & Diamond, M. L. (2023). Per- and polyfluoroalkyl substances in canadian fast food packaging. *Environmental Science & Technology Letters*, 10(4), 343–349. https://doi.org/10.1021/acs.estlett.2c00926.

Sciscione, F., Hailes, H. C., & Miodownik, M. (2023). The performance and environmental impact of pro-oxidant additive containing plastics in the open unmanaged environment—A review of the evidence. *Royal Society Open Science*, 10(5), 230089. https://doi.org/10.1098/rsos.230089.

Selvamurugan Muthusamy, M., & Pramasivam, S. (2019). Bioplastics – an eco-friendly alternative to petrochemical plastics. *Current World Environment*, 14(1), 49–59. https://doi.org/10.12944/CWE.14.1.07.

Shaw, M. (n.d.). Plastics in the furniture industry. *Plastics in Australia*, January.

Shove, E. (Ed.). (2007). *The design of Everyday life*. Berg.

Shove, E., & Walker, G. (2007). Caution! Transitions ahead: Politics, practice, and sustainable transition management. *Environment and Planning*. A, 39(4), 763–770. https://doi.org/10.1068/a39310.

Silveira, S., & Johnson, F. X. (2016). Navigating the transition to sustainable bioenergy in Sweden and Brazil: Lessons learned in a European and International context. *Energy Research & Social Science*, 13, 180–193. https://doi.org/10.1016/j.erss.2015.12.021.

Simon, N., Raubenheimer, K., Urho, N., Unger, S., Azoulay, D., Farrelly, T., Sousa, J., Asselt, H. van, Carlini, G., Sekomo, C., Schulte, M. L., Busch, P.-O., Wienrich, N., & Weiand, L. (2021). A binding global agreement to address the life cycle of plastics. *Science*, 373(6550), 43–47. https://doi.org/10.1126/science.abi9010.

Simona, S., Oliviero, M., Salvatore, C., & Manzo, S. (2020). Adverse effects of oxo-degradable plastic leachates in freshwater environment. *Environmental Science and Pollution Research*, 27(8), 8586–8595. https://doi.org/10.1007/s11356-019-07466-z.

Sin, L. T., & Tuen, B. S. (2022). *Plastics and sustainability* (1st ed.). Elsevier.

Smink, M. M., Hekkert, M. P., & Negro, S. O. (2015). Keeping sustainable innovation on a leash? Exploring incumbents' institutional strategies. *Business Strategy & the Environment (John Wiley & Sons, Inc)*, 24(2), 86–101. https://doi.org/10.1002/bse.1808.

Smith, A. (2003). Transforming technological regimes for sustainable development: A role for alternative technology niches? *Science & Public Policy (SPP)*, 30(2), 127. https://doi.org/10.3152/147154303781780623.

Smith, A. (2007). Translating sustainabilities between green niches and socio-technical regimes. *Technology Analysis & Strategic Management*, 19(4), 427–450. https://doi.org/10.1080/09537320701403334.

Smith, A., & Raven, R. (2012). What is protective space? Reconsidering niches in transitions to sustainability. *Research Policy*, 41(6), 1025–1036. https://doi.org/10.1016/j.respol.2011.12.012.

Smith, A., Stirling, A., & Berkhout, F. (2005). The governance of sustainable socio-technical transitions. *Research Policy*, 34(10), 1491–1510. https://doi.org/10.1016/j.respol.2005.07.005.

Smithsonian. (2012). *The age of plastic: Ingenuity and responsibility*. 2012 MCI Symposium, Washington, DC.

Spalding, M. A., & Chatterjee, A. (2017). *Handbook of industrial polyethylene and technology: Definitive guide to manufacturing, properties, processing, applications and markets set*. John Wiley & Sons, Inc.

Sparke, P. (1997). Review of American plastic: A cultural history [review of *Review of American plastic: A cultural history*, by J. L. Meikle]. *Studies in the Decorative Arts*, 4(2), 110–112.

Sparke, P., Mossman, S. T. I., Friedel, R., & Victoria and Albert Museum (Eds.). (1993). *The plastics age: From modernity to post-modernity* (Repr. of the 1990 ed.). Victoria & Albert Pubs.

Stafford, R., & Jones, P. J. S. (2019). Viewpoint – Ocean plastic pollution: A convenient but distracting truth? *Marine Policy*, 103, 187–191. https://doi.org/10.1016/j.marpol.2019.02.003

Statista. (2022, July 27). Italy: Furniture company Felofin SpA sales. https://www.statista.com/statistics/977696/sales-value-of-furniture-company-felofin-spa (accessed 9 March 2024).

Storace, E., Holzwarth, H. W., Annicchiarico, S., Dorfles, G., Hamaide, C., Jousset, M.-L., Miller, R. C., Odoni, G., Sozzani, F., Sudjic, D., Crivelli, G., Luti, C., & Kartell (Firm) (Eds.). (2012). *The culture of plastics*. Taschen GmbH.

Storrow, B. (2020, 24 January). *Plastics plants are poised to be the next big carbon superpolluters*. Scientific American. https://www.scientificamerican.com/article/plastics-plants-are-poised-to-be-the-next-big-carbon-superpolluters (accessed 9 March 2024).

Streb, J. (2003). Shaping the national system of inter-industry knowledge exchange: Vertical integration, licensing and repeated knowledge transfer in the German plastics industry. *Research Policy*, 32(6), 1125–1140. https://doi.org/10.1016/S0048-7333(02)00114-2.

Suzdaltsev, J. (2015, 28 January). White plastic chairs are taking over the world. *Vice*. https://www.vice.com/en_us/article/bn5e4m/white-plastic-chairs-are-taking-over-the-world-128 (accessed 9 March 2024).

Swan, S. H., & Colino, S. (2021). *Count down: How our modern world is threatening sperm counts, altering male and female reproductive development, and imperiling the future of the human race* (First Scribner hardcover ed.). Scribner.

Swedlow. (1941). *Plastics by Swedlow* (company brochure). Swedlow.

Taft, M. L. (2014). *Making Danish modern, 1945–1960*. PhD, University of Chicago.

The Age of Plastics. (1950, 7 July). *Age*, 3.

The Australian Plastics Industry. (1950, 24 April). *Building and Engineering*, 4, 61.

The Bars Go Up. (1940). *TIME Magazine*, 36(3), 60–62.

The Business Research Company. (2023a). *Global transportation manufacturing market report*. https://www.thebusinessresearchcompany.com/report/transportation-manufacturing-global-market-report (accessed 9 March 2024).

The Business Research Company. (2023b). *Oil & gas upstream activities market opportunities, growth drivers 2032*. https://www.thebusinessresearchcompany.com/report/oil-and-gas-upstream-activities-global-market-report (accessed 9 March 2024).

The Business Research Company. (2023c). *Petrochemicals global market growth demand, strategies, outlook By 2032*. https://www.thebusinessresearchcompany.com/report/petrochemicals-global-market-report (accessed 9 March 2024).

The Business Research Company. (2023d). *Waste management and remediation services market size, growth and forecast 2032*. https://www.thebusinessresearchcompany.com/report/waste-management-and-remediation-services-global-market-report (accessed 9 March 2024).

The Henry Ford Museum. (n.d.). Soybean Car—The Henry Ford. https://www.thehenryford.org/collections-and-research/digital-resources/popular-topics/soy-bean-car/ (accessed 9 March 2024).

The presidency the last step taken. (1941). *TIME Magazine*, 38(5), 11–12.

The prospects for plastics. (1974). *Plastics News*, August.

Thompson, R. C. (2013). Plastics, environment and health. In *Accumulation: The material politics of plastic* (pp. 150–168). Routledge, Taylor & Francis Group.

Thorpe, A. (2010). Design's role in sustainable consumption. *Design Issues*, 26(2), 3–16. JSTOR.

Together for Sustainability. (2022). *The product carbon footprint guideline for the chemical industry*. https://www.tfs-initiative.com (accessed 5 October 2022).

Tonkinwise, C., Kossoff, G., & Irwin, T. (2015). Transition design provocation. *Design Philosophy Papers*, 13(1), 3–11.

Tribbe, M. D. (2014). *No requiem for the space age: The Apollo moon landings and American culture*. Oxford University Press.

Trimble, A. (2022, 14 September). Louisiana court vacates air permits for Formosa's massive petrochemical complex in Cancer Alley. Earthjustice. https://earthjustice.org/news/press/2022/louisiana-court-vacates-air-permits-for-formosas-massive-petrochemical-complex-in-cancer-alley (accessed 9 March 2024).

Tullo, A. (2021, 11 January). Europe is implementing a tax on plastic. *C&EN Global Enterprise*, 99(2), 34–34. https://doi.org/10.1021/cen-09902-cover7.

UN News. (2021, 2 March). Environmental racism in Louisiana's 'Cancer Alley', must end, say UN human rights experts. UN News. https://news.un.org/en/story/2021/03/1086172 (accessed 9 March 2024).

United Nations Environment Programme. (2024). *Global waste management outlook 2024*. http://www.unep.org/resources/global-waste-management-outlook-2024 (accessed 9 March 2024).

US Energy Information Administration. (2022). U.S. crude oil first purchase price (dollars per barrel). https://www.eia.gov/dnav/pet/hist/LeafHandler.ashx?n=PET&s=F000000__3&f=A (accessed 9 March 2024).

Van Doren, H. (1940). *Industrial Design*. McGraw Hill.

van Mossel, A., van Rijnsoever, F. J., & Hekkert, M. P. (2018). Navigators through the storm: A review of organization theories and the behavior of incumbent firms during transitions. *Environmental Innovation and Societal Transitions*, 26, 44–63. https://doi.org/10.1016/j.eist.2017.07.001.

van Waes, A., Farla, J., Frenken, K., de Jong, J. P. J., & Raven, R. (2018). Business model innovation and socio-technical transitions. A new prospective framework with an application to bike sharing. *Journal of Cleaner Production*, 195, 1300–1312. https://doi.org/10.1016/j.jclepro.2018.05.223.

Verbruggen, A. (2022a). The geopolitics of trillion US$ oil & gas rents. *International Journal of Sustainable Energy Planning and Management*, 36, 3–10. https://doi.org/10.54337/ijsepm.7395.

Verbruggen, A. (2022b, 15 October). Quo Vadis energy system transformation? https://doi.org/10.13140/RG.2.2.13975.04007.

Vico Magistretti – Heller. (n.d.). http://hellerinc.com/designers/vico-magistretti (accessed 5 October 2022).

Victoria and Albert Museum, & Whitechapel Art Gallery (Eds.). (1971). *Modern chairs, 1918–1970*. Boston Book and Art.

Vidal, J. (2020, 1 February). The plastic polluters won 2019 – and we're running out of time to stop them, *The Guardian*. https://www.theguardian.com/environment/2020/jan/02/year-plastic-pollution-clean-beaches-seas?CMP=Share_iOSApp_Other (accessed 9 March 2024).

Vogel, S. A. (2012). *Is it safe?: BPA and the struggle to define the safety of chemicals*. University of California Press.

Vondom. (2021, 29 June). Vondom revolution | Sustainable design. https://www.vondom.com/revolution (accessed 9 March 2024).

Waage, S. (2007). Re-considering product design: A practical '"road-map"' for integration of sustainability issues. *Journal of Cleaner Production*, 15, 638–649.

Wainstein, M. E., & Bumpus, A. G. (2016). Business models as drivers of the low carbon power system transition: A multi-level perspective. *Journal of Cleaner Production*, 126, 572–585. https://doi.org/10.1016/j.jclepro.2016.02.095.

Watson, A. (2002). *Mod to Memphis: Design in colour 1960s–80s*. Powerhouse Pub.

Watson, R. T. (Ed.). (2001). *IPCC – Climate change 2001: Synthesis report | GRID-Arendal*. Intergovernmental Panel on Climate Change. https://www.grida.no/publications/267 (accessed 9 March 2024).

Wellenreuther, C., & Wolf, A. (2020). *Innovative feedstocks in biodegradable bio-based plastics: A literature review*. Hamburg Institute of International Economics.

Westermann, A. (2013). When consumer citizens spoke up: West Germany's early dealings with plastic waste. *Contemporary European History*, 22(3), 477–498. http://dx.doi.org/10.1017/S0960777313000246.

White, K., Hardisty, D. J., & Habib, R. (2019, 1 July). The elusive green consumer. *Harvard Business Review*, July–August, 124–133.

Wingfield, P. (2018). Starck biography. Starck. https://www.starck.com/about (accessed 9 March 2024).

Worsoe, E. (1974). The plastic revolution. *Australian House and Garden*, February.

WRAP UK. (2010). *Environmental benefits of recycling*. https://www.wrap.ngo/resources/report/environmental-benefits-recycling-2010-update (accessed 5 October 2022).

Wright, Y. (1951). Let's look at plastics. *Australian House and Garden*, February.

Xu, E. G., & Ren, Z. J. (2021). Preventing masks from becoming the next plastic problem. *Frontiers of Environmental Science & Engineering*, 15(6), 125. https://doi.org/10.1007/s11783-021-1413-7.

Yang, Y., Xie, E., Du, Z., Peng, Z., Han, Z., Li, L., Zhao, R., Qin, Y., Xue, M., Li, F., Hua, K., & Yang, X. (2023). Detection of various microplastics in patients undergoing cardiac surgery. *Environmental Science & Technology*, 57(30), 10911–10918. https://doi.org/10.1021/acs.est.2c07179.

Yarsley, V., & Couzens, E. (1941). *Plastics*. Penguin.

Yarsley, V., & Couzens, E. (1968). *Plastics in the modern world*. Penguin.

Yin, S., Tuladhar, R., Shi, F., Shanks, R. A., Combe, M., & Collister, T. (2015). Mechanical reprocessing of polyolefin waste: A review. *Polymer Engineering & Science*, 55(12), 2899–2909. https://doi.org/10.1002/pen.24182.

Young America Films (director). (1945, January). *History of plastics: Plastics in World War II: Plastics (1944) – CharlieDeanArchives*, YouTube. https://www.youtube.com/watch?v=GirvOmjPZrc (accessed 9 March 2024).

Zalasiewicz, J., Waters, C. N., Ivar Do Sul, J. A., Corcoran, P. L., Barnosky, A. D., Cearreta, A., Edgeworth, M., Gałuszka, A., Jeandel, C., Leinfelder, R., McNeill, J. R., Steffen, W., Summerhayes, C., Wagreich, M., Williams, M., Wolfe, A. P., & Yonan, Y. (2016). The geological cycle of plastics and their use as a stratigraphic indicator of the Anthropocene. *Anthropocene*, 13, 4–17. https://doi.org/10.1016/j.ancene.2016.01.002.

Zheng, J., & Suh, S. (2019). Strategies to reduce the global carbon footprint of plastics. *Nature Climate Change*, 9, 374–378.

Zhuang, J. (2018). The chair that's everywhere by Justin Zhuang. *Works That Work Magazine*. https://worksthatwork.com/10/the-chair-thats-everywhere (accessed 9 March 2024).

Zuckerman, E. (2011, 4 June). Those white plastic chairs – the monobloc and the context-free object | … My heart's in Accra. http://www.ethanzuckerman.com/blog/2011/04/06/those-white-plastic-chairs-the-monobloc-and-the-context-free-object (accessed 9 March 2024).

Index